introduction to colloid and surface chemistry

D J Shaw

Second Edition

Butterworths

Introduction to Colloid
and Surface Chemistry

Introduction to Colloid and Surface Chemistry

DUNCAN J. SHAW
Department of Chemistry, Liverpool Polytechnic

¼ sep
CHEM

BUTTERWORTHS
LONDON - BOSTON
Sydney - Wellington - Durban - Toronto

United Kingdom   Butterworth & Co. (Publishers) Ltd
London   88 Kingsway, WC2B 6AB

Australia   Butterworths Pty Ltd
Sydney   586 Pacific Highway, Chatswood, NSW 2067
Also at Melbourne, Brisbane, Adelaide and Perth

South Africa   Butterworth & Co. (South Africa) (Pty) Ltd
Durban   152−154 Gale Street

New Zealand   Butterworths of New Zealand Ltd
Wellington   26−28 Waring Taylor Street, 1

Canada   Butterworth & Co. (Canada) Ltd
Toronto   2265 Midland Avenue,
Scarborough, Ontario, M1P 4S1

USA   Butterworth (Publishers) Inc
Boston   19 Cummings Park, Woburn, Massachusetts, 01801

First Published 1966
Second Impression (Revised) 1968
Second Edition 1970
Second Impression 1975
Third Impression 1976

ISBN 0 408 70021 1

Suggested UDC number: 541·182/·183

Filmset and Printed in England by
Cox & Wyman Ltd., Fakenham

To Ann

Preface
to the Second Edition

In preparing this Second Edition, the main structure of the book has not been altered. A number of minor changes and updatings have been made throughout the text. The major changes from the First Edition consist of the addition of sections on Adsorption from Solution (Chapter 6) and Foams (Chapter 10), and a remodelling and expansion of Chapters 7 and 8 on Charged Interfaces and Colloid Stability. The text has been modified to conform with the International System of Units (SI) and recently proposed physico-chemical methods of expression.

D. J. SHAW

Liverpool

Preface
to the First Edition

I have felt for some time the need for a textbook on colloid and surface chemistry offering a standard and overall coverage inter-mediate between those found in the appropriate chapters of a general textbook on physical chemistry and in the more comprehensive and specialised treatises on colloid and/or surface chemistry. It is in an attempt to fill this gap in the literature that this book has been written.

In writing the book, I have kept a number of audiences in mind, particularly: university and college of technology students studying for an honours degree or its equivalent, or commencing a programme of postgraduate research; scientists in industry who desire a broad background in a subject which may have been somewhat neglected during academic training; and those interested in branches of natural science, for whom an understanding of colloid and surface pheno-mena is essential.

The subject matter is generally approached from a fundamental angle, and the reader is assumed to possess a knowledge of the basic principles of physical chemistry. Opportunities have also been taken to describe many of the practical applications of this subject. In addition, a list of references for further reading, mainly books and review articles, is given at the end of the book.

I would like to thank Dr A. L. Smith, Dr F. MacRitchie and Mr A. M. Shaw for their many helpful suggestions relating to the preparation of this book.

D. J. SHAW

Liverpool

ix

Contents

PREFACE TO SECOND EDITION   vii
PREFACE TO FIRST EDITION   ix

1. THE COLLOIDAL STATE   1
   Introduction   1
   Classification of Colloidal Systems   2
   Structural Characteristics   5
   Preparation and Purification of Colloidal Systems   8

2. KINETIC PROPERTIES   16
   The Motion of Particles in Liquid Media   16
   Brownian Motion and Translational Diffusion   18
   The Ultracentrifuge   26
   Osmotic Pressure   31
   Rotary Brownian Motion   39

3. OPTICAL PROPERTIES   41
   Light Scattering   41
   Electron Microscopy and Dark-Field Microscopy   48

4. LIQUID–GAS AND LIQUID–LIQUID INTERFACES   55
   Surface and Interfacial Tensions   55
   Adsorption and Orientation at Interfaces   64

Association Colloids 71
Spreading 77
Monomolecular Films 80

5. THE SOLID–GAS INTERFACE 98
Adsorption of Gases and Vapours on Solids 98

6. THE SOLID–LIQUID INTERFACE 117
Contact Angles and Wetting 117
Ore Flotation 122
Detergency 124
Adsorption from Solution 128

7. CHARGED INTERFACES 133
The Electric Double Layer 133
Electrokinetic Phenomena 147
Electrokinetic Theory 156

8. COLLOID STABILITY 167
Lyophobic Sols 167
Systems Containing Lyophilic Material 183

9. RHEOLOGY 187
Introduction 187
Viscosity 188
Non-Newtonian Flow 196
Viscoelasticity 199

10. EMULSIONS AND FOAMS 206
Oil in Water and Water in Oil Emulsions 206
Emulsion Polymerisation 211
Foams 213

PROBLEMS 219

ANSWERS 226

BIBLIOGRAPHY 228

INDEX 233

1

The Colloidal State

INTRODUCTION

Colloid science concerns systems in which one or more of the components has at least one dimension within the range of about 1 nm to 1 μm;* i.e. it concerns, in the main, systems containing either large molecules or small particles. The adjective 'microhetero-geneous' provides an appropriate description of most colloidal systems. There is, however, no sharp distinction between colloidal and non-colloidal systems, especially at the upper size limit; for example, the droplet size in emulsions is normally in excess of 1 μm, yet it is convenient to treat them as colloidal systems.

In many respects colloid science links together some of the purer scientific disciplines. Particularly important is the application of physicochemical techniques to the study of natural systems, especially the proteins. The field of synthetic high polymers is another notable offshoot of colloid science. Colloidal phenomena are very frequently encountered in industrial practice: plastics, rubber, paint, deter-gents, paper, soil, foodstuffs, fabrics, precipitation, chromatography, ion exchange, flotation and heterogeneous catalysis are just a few examples of materials and techniques which deal with matter of colloidal dimensions. Owing to the immense complexity of colloidal systems the subject cannot often be treated with the exactness

* 1 nm = 10^{-9} m;   1 μm = 10^{-6} m.

associated with certain branches of physical chemistry, and it is this lack of precision rather than a lack of importance which is probably responsible for an unjustifiable tendency to neglect colloid science during academic training.

Until the last few decades colloid science stood more or less on its own as an almost entirely descriptive subject which did not appear to fit within the general framework of physics and chemistry. The use of materials of doubtful composition, which put considerable strain on the questions of reproducibility and interpretation, was partly responsible for this state of affairs. Nowadays, the tendency is to work whenever possible with pure materials, which act as models for the actual systems under consideration. McBain's work on soaps and detergents typifies such an approach. Despite the large number of variables which are often involved, research of this nature coupled with advances in the understanding of the fundamental principles of physics and chemistry has made it possible to formulate coherent, if not comprehensive, theories relating to many of the aspects of colloidal behaviour.

The natural laws of physics and chemistry which describe the behaviour of matter in the massive and molecular states can also be applied to the colloidal state. The characteristic feature of colloid science lies in the relative importance which is attached to the various physicochemical properties of the systems being studied. As we shall see, the factors which contribute most to the overall nature of a colloidal system are:

> Particle size
> Particle shape and flexibility
> Surface (including electrical) properties
> Particle–particle interactions
> Particle–solvent interactions

CLASSIFICATION OF COLLOIDAL SYSTEMS

Colloidal systems may be grouped into three general classifications:

1. Colloidal dispersions: These are thermodynamically unstable owing to their high surface free energy and are irreversible systems in the sense that they are not easily reconstituted after phase separation.

2. True solutions of macromolecular material (natural or syn-

thetic): These are thermodynamically stable and reversible in the sense that they are easily reconstituted after separation of solute from solvent.

3. Association colloids (sometimes referred to as colloidal electrolytes), which are thermodynamically stable (see Chapter 4).

Dispersions

The particles in a colloidal dispersion are sufficiently large for definite surfaces of separation to exist between the particles and the medium in which they are dispersed. Simple colloidal dispersions are, therefore, two-phase systems. The phases are distinguished by the terms *dispersed phase* (for the phase forming the particles) and *dispersion medium* (for the medium in which the particles are distributed)—see Table 1.1. The physical nature of a dispersion depends, of course, on the respective roles of the constituent phases; for example, an oil in water (O/W) emulsion and a water in oil (W/O) emulsion could have almost the same overall composition, but their physical properties would be notably different (see Chapter 10).

Table 1.1   TYPES OF COLLOIDAL DISPERSION

Dispersed phase	Dispersion medium	Name	Examples
Liquid	Gas	Liquid aerosol	Fog, liquid sprays
Solid	Gas	Solid aerosol	Smoke, dust
Gas	Liquid	Foam	Foam on soap solutions, fire-extinguisher foam
Liquid	Liquid	Emulsion	Milk, mayonnaise
Solid	Liquid	Sol, colloidal suspension; Paste (high solid concentration)	Au sol, AgI sol; toothpaste
Gas	Solid	Solid foam	Expanded polystyrene
Liquid	Solid	Solid emulsion	Opal, pearl
Solid	Solid	Solid suspension	Pigmented plastics

Sols and emulsions are by far the most important types of colloidal dispersion. The term *sol* is used to distinguish colloidal suspensions from macroscopic suspensions; there is, of course, no sharp line of demarcation. When the dispersion medium is aqueous the term *hydrosol* is generally used. Foams are somewhat different in that it is the dispersion medium which has colloidal dimensions.

The importance of the interface

The essential character common to all colloidal dispersions is the large area-to-volume ratio for the particles involved. At the inter-faces between the dispersed phase and the dispersion medium char-acteristic surface properties such as adsorption and electrical double layer effects are evident and play a very important part in determining the physical properties of the system as a whole. For this reason, surface chemistry is closely linked with colloid science.

The surface or interfacial phenomena associated with colloidal systems such as emulsions and foams are often studied by means of experiments on artificially prepared flat surfaces rather than on the colloidal systems themselves. Such methods provide a most useful indirect approach to the various problems involved.

Lyophilic and lyophobic systems

The terms *lyophilic* (liquid loving) and *lyophobic* (liquid hating) are frequently used to describe the tendency of a surface or func-tional group to become wetted or solvated. If the liquid medium is aqueous the terms *hydrophilic* and *hydrophobic* are used.

Lyophilic surfaces can be made lyophobic and vice versa. For example, clean glass surfaces, which are hydrophilic, can be made hydrophobic by a coating of wax; conversely, the droplets in a hydrocarbon oil in water emulsion, which are hydrophobic, can be made hydrophilic by the addition of protein to the emulsion, the protein molecules adsorbing on to the droplet surfaces.

This terminology is particularly useful when considering the phenomenon of surface activity. The molecules of surface-active materials have a strong affinity for interfaces because they contain both hydrophilic and lipophilic (oil loving) regions.

The general usage of the terms lyophilic and lyophobic in describing colloidal systems is somewhat illogical. Lyophobic traditionally describes liquid dispersions of solid or liquid particles produced by mechanical or chemical action; however, in these so-called 'lyophobic sols' (e.g. dispersions of powdered alumina or silica in water) there is often a high affinity between the particles and the dispersion medium, i.e. the particles are really lyophilic. Indeed, if the term lyophobic is taken to imply no affinity between particles and dispersion medium (an unreal situation) then the particles would not be wetted and no dispersion could, in fact, be formed. Lyophilic traditionally describes soluble macromolecular material; however, lyophobic regions are often present, for example, proteins are partly hydrophobic (hydrocarbon regions) and partly hydrophilic (peptide linkages, amino and carboxyl groups).

STRUCTURAL CHARACTERISTICS

Particle shape

Particle asymmetry is a factor of considerable importance in determining the overall properties (especially those of a mechanical nature) of colloidal systems. Roughly speaking, colloidal particles can be classified according to shape as *corpuscular, laminar* or *linear* (see, for example, the electron micrographs in Figure 3.6). The exact shape may be complex but, to a first approximation, the

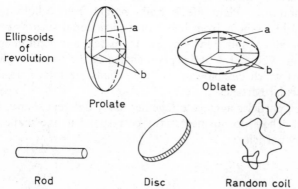

Figure 1.1. Some model representations for non-spherical particles

particles can often be treated theoretically in terms of models which have relatively simple shapes (Figure 1.1).

The easiest model to treat theoretically is the sphere, and many colloidal systems do, in fact, contain spherical or nearly spherical particles. Emulsions, latexes (dispersions of polymeric materials such as rubbers and plastics in water), liquid aerosols, etc., contain spherical particles. Certain protein molecules are approximately spherical. The crystallite particles in dispersions such as gold sol are sufficiently symmetrical to behave like spheres.

Corpuscular particles which deviate from spherical shape can often be treated theoretically as ellipsoids of revolution. Many proteins approximate this shape. An ellipsoid of revolution is characterised by its axial ratio, which is the ratio of the single half-axis a to the radius of revolution b. The axial ratio is greater than unity for a prolate (rugby football-shaped) ellipsoid, and less than unity for an oblate (discus-shaped) ellipsoid.

Ferric oxide and clay suspensions are examples of systems containing plate-like particles.

High-polymeric material usually exists in the form of long thread-like straight or branched-chain molecules. As a result of inter-chain attraction or cross-linking (arising from covalent bonding, hydrogen bonding or van der Waals forces) and entanglement of the polymer chains, these materials often exhibit considerable mechanical strength and durability. This is not possible when the particles are corpuscular or laminar.

In nature, thread-like polymeric material fulfils an essential structural role. Plant life is built mainly from cellulose fibres. Animal life is built from linear protein material such as collagen in skin, sinew and bone, myosin in muscle and keratin in nails and hair. The coiled polypeptide chains of the so-called globular proteins which circulate in the body fluids are folded up to give corpuscular particles.

When particles aggregate together many different shapes can be formed. These do not necessarily correspond to the shape of the primary particles.

Flexibility

Thread-like high-polymer molecules show considerable flexibility due to rotation about carbon–carbon and other bonds. In solution, the shape of these molecules alters continuously under the influence of thermal motion and a rigid rod model is therefore unsuitable. A better theoretical treatment is to consider the polymer molecules as random coils, but even this model is not completely accurate.

Rotation about bonds does not permit complete flexibility, and steric and excluded volume effects also oppose the formation of a truly random configuration, so that, in these respects, dissolved linear polymer molecules will tend to be more extended than random coils. The relative magnitudes of polymer–polymer and polymer–solvent forces must also be taken into account. If the segments of the polymer chain tend to stick to one another then a tighter than random coil, and possibly precipitation, will result; whereas a looser coil results when the polymer segments tend to avoid one another because of strong solvation and/or electrical repulsion.

Solvation

Colloidal particles are usually solvated, often to the extent of about one molecular layer and this tightly bound solvent must be treated as a part of the particle.

Sometimes much greater amounts of solvent can be immobilised by mechanical entrapment within particle aggregates. This occurs when voluminous flocculent hydroxide precipitates are formed. In solutions of long thread-like molecules the polymer chains may cross-link, chemically or physically, and/or become mechanically entangled to such an extent that a continuous three-dimensional network is formed. If all of the solvent becomes mechanically trapped and immobilised within this network, the system as a whole takes on a solid appearance and is called a *gel*.

Polydispersity and the averages

The terms *relative molecular mass** and *particle size* can only have well-defined meanings when the system under consideration is *monodispersed,* i.e. when the molecules or particles are all alike. Colloidal systems are generally of a *polydispersed* nature, i.e. the particles in a particular sample vary in size. Since a detailed determination of relative molecular mass or particle size distribution is often impracticable, less perfect experimental methods, which yield average values, must be accepted. The significance of the word *average* depends on the relative contributions of the various mole-

* Formerly called molecular weight.[1]

cules or particles to the property of the system which is being measured.

Osmotic pressure, which is a colligative property, depends simply on the number of solute molecules present and so yields a *number-average* relative molecular mass:

$$M_r \text{ (number average)} = \frac{\Sigma n_i M_{r,i}}{\Sigma n_i} \qquad \ldots (1.1)$$

where n_i is the number of molecules of relative molecular mass $M_{r,i}$.

In most cases, the larger particles make a greater individual contribution to the property being measured. If the contribution of each particle is proportional to its mass (as in light scattering) a *mass-average* relative molecular mass or particle mass is given:

$$M_r \text{ (mass average)} = \frac{\Sigma n_i M_{r,i}^2}{\Sigma n_i M_{r,i}} \qquad \ldots (1.2)$$

For any polydispersed system, M_r (mass average) $> M_r$ (number average), and only when the system is monodispersed will these averages coincide. The ratio, M_r (mass average)$/M_r$ (number average), is a measure of the degree of polydispersity.

PREPARATION AND PURIFICATION OF COLLOIDAL SYSTEMS

Colloidal dispersions

Basically, the formation of colloidal material involves either degradation of bulk matter or aggregation of small molecules or ions.

Dispersion of bulk material by simple grinding in a colloid mill or by ultrasonics does not generally lead to extensive subdivision owing to the tendency of smaller particles to reunite (*a*) under the influence of the mechanical forces involved and (*b*) by virtue of the attractive forces between the particles. After prolonged grinding the distribution of particle sizes reaches an equilibrium. Somewhat finer dispersions can be obtained by incorporating an inert diluent to reduce the chances of the particles in question encountering one another during the grinding, or by wet-milling in the presence of surface-active material. As an example of the first of these techniques, a sulphur sol in the upper colloidal range can be prepared by grinding a mixture of sulphur and glucose, dispersing the resulting powder in water and then removing the dissolved glucose from the sol by dialysis.

A higher degree of dispersion is generally obtainable when a sol is prepared by an aggregation method. Aggregation methods involve the formation of a molecularly dispersed supersaturated solution from which the material in question precipitates in a suitably divided form. A variety of methods, such as the substitution of a poor solvent for a good one, cooling and various chemical reactions, can be utilized to achieve this end.

A coarse sulphur sol can be prepared by pouring a saturated solution of sulphur in alcohol or acetone into water just below boiling point. The alcohol or acetone vaporises, leaving the water-insoluble sulphur colloidally dispersed. This technique is convenient for dispersing wax-like material in an aqueous medium.

Examples of chemical reactions which yield sols under suitably controlled experimental conditions are

$$FeCl_3 + H_2O \text{ (boiling)} \rightarrow Fe_2O_3 \text{ (hydrated)}$$
$$AgNO_3 + KI \rightarrow AgI \text{ (see reference 16)}$$
$$Na_2S_2O_3 + HCl \rightarrow S \text{ (see reference 17)}$$
$$HAuCl_4 + HCHO \rightarrow Au$$

Nucleation and growth

The formation of a new phase during precipitation involves two distinct stages—*nucleation* (the formation of centres of crystallisation) and crystal growth—and (leaving aside the question of stability) it is the relative rates of these processes which determine the particle size of the precipitate so formed.[18] A high degree of dispersion is obtained when the rate of nucleation is high and the rate of crystal growth is low.

The initial rate of nucleation depends on the degree of super-saturation which can be reached before phase separation occurs, so that colloidal sols are most easily prepared when the substance in question has a very low solubility. With material as soluble as, for example, calcium carbonate, there is a tendency for the smaller particles to redissolve (see page 57) and recrystallise on the larger particles as the precipitate is allowed to age.

The rate of particle growth depends mainly on the following factors:

1. The amount of material available.
2. The viscosity of the medium, which controls the rate of diffusion of material to the particle surface.
3. The ease with which the material is correctly orientated and incorporated into the crystal lattice of the particle.
4. Adsorption of impurities on the particle surface, which act as growth inhibitors.
5. Particle–particle aggregation.

Von Weimarn (1908) investigated the dependence on reagent concentration of the particle sizes of barium sulphate precipitates formed in alcohol–water mixtures by the reaction

$$Ba(CNS)_2 + MgSO_4 \rightarrow BaSO_4 + Mg(CNS)_2.$$

At very low concentrations, *ca.* 10^{-4} to 10^{-3} mol dm^{-3}, the supersaturation is sufficient for extensive nucleation to occur, but crystal growth is limited by the availability of material, with the result that a sol is formed. At moderate concentrations, *ca.* 10^{-2} to 10^{-1} mol dm^{-3}, the extent of nucleation is not much greater, so that more

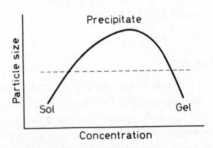

Figure 1.2. The dependence of particle size on reagent concentration for the precipitation of a sparingly soluble material

material is available for crystal growth and a coarse filterable precipitate is formed. At very high concentrations, *ca.* 2 to 3 mol dm^{-3}, the high viscosity of the medium slows down the rate of crystal growth sufficiently to allow time for much more extensive nucleation and the formation of very many small particles. Owing to their closeness, the barium sulphate particles will tend to link and the dispersion will take the form of a translucent, semi-solid gel.

Monodispersed sols

Aggregation methods generally lead to the formation of polydispersed sols, mainly because the formation of new nuclei and the growth of established nuclei occur simultaneously, and so the particles finally formed are grown from nuclei formed at different times. In experiments designed to test the validity of theories, however, there are obvious advantages attached to the use of monodispersed systems. The preparation of such systems requires conditions where nucleation is restricted to a relatively short period at the start of the sol formation. This situation can sometimes be achieved either by seeding a supersaturated solution with very small particles, or under conditions which lead to a short burst of homogeneous nucleation.

An example of the seeding technique is based on that of Zsigmondy (1906) for preparing approximately monodispersed gold sols. A hot dilute aqueous solution of $HAuCl_4$ is neutralised with potassium carbonate and a part of the solute is reduced with a small amount of white phosphorus to give a highly dispersed gold sol with an average particle radius of *ca.* 1 nm. The remainder of the $HAuCl_4$ is then reduced relatively slowly with formaldehyde in the presence of these small gold particles. Further nucleation is thus effectively avoided and all of the gold produced in this second stage accumulates on the seed particles. Since the absolute differences in the seed particle sizes are not great, an approximately monodispersed sol is formed. By regulating the amount of $HAuCl_4$ reduced in the second stage and the number of seed particles produced in the first stage, the gold particles can be grown to a desired size.

A similar seeding technique can be used to prepare monodispersed polystyrene latex dispersions by emulsion polymerisation (see page 212).

Among the monodispersed sols which have been prepared under conditions which lead to a short burst of homogeneous nucleation are (*a*) sulphur sols,[17] formed by mixing very dilute aqueous solutions of HCl and $Na_2S_2O_3$, (*b*) silver bromide sols,[16] by controlled cooling of hot saturated aqueous solutions of silver bromide, and (*c*) silver bromide and silver iodide sols,[16] by diluting aqueous solutions of the complexes formed in the presence of excess silver or halide ions. In each case, the concentration of the material of the dispersed phase slowly passes the saturation point and attains a degree of supersaturation at which nucleation becomes appreciable. Since the generation of dispersed phase material in excess of the saturation concentration is slow, the appearance of nuclei and the accompanying relief of supersaturation is restricted to a relatively short period and few new nuclei are formed after this initial outburst. The nuclei then grow uniformly by a diffusion-controlled process and a sol of monodispersed particles is formed.

Macromolecular colloids

Macromolecular chemistry covers a particularly wide field which includes natural polymeric material, such as proteins, cellulose, gums and natural rubber; industrial derivatives of natural polymers, such as sodium carboxymethyl cellulose, rayon and vulcanised rubber; and the purely synthetic polymers, such as polythene (polyethylene), teflon (polytetrafluoroethylene), polystyrene, perspex (polymethyl methacrylate), terylene (polyethylene terephthalate) and the nylons (e.g. polyhexamethylene adipamide). Only brief mention of some of the more general aspects of polymerisation will be made. The reader is referred to the various specialised texts for details of preparation, properties and utilisation of these products.

High polymers contain giant molecules which are built up from a large number of similar (but not necessarily identical) units (or monomers) linked by primary valence bonds. Polymerisation reactions can be performed either in the bulk of the monomer material or in solution. A further technique, emulsion polymerisation, which permits far greater control over the reaction, is discussed in Chapter 10.

There are two distinct types of polymerisation: addition polymerisation and condensation polymerisation.

Addition polymerisation does not involve a change of chemical composition. It generally proceeds by a chain mechanism, a typical series of reactions being:

1. Formation of free radicals from a catalyst (initiator), such as a peroxide.
2. Initiation; for example

$$\underset{\substack{\text{vinyl}\\\text{monomer}}}{CH_2 = CHX} + \underset{\substack{\text{free}\\\text{radical}}}{R\bigstar} \rightarrow RCH_2 - \overset{\bigstar}{C}HX$$

3. Propagation

$$RCH_2 - \overset{\star}{C}HX + CH_2 = CHX \rightarrow$$

$$RCH_2 - CHX - CH_2 - \overset{\star}{C}HX, \text{ etc.,}$$

to

$$R\underset{\text{vinyl polymer}}{(CH_2 - CHX)_n CH_2 - \overset{\star}{C}HX}$$

4. Termination. This can take place in several ways, such as reaction of the activated chain with an impurity, an additive or other activated chains, or by disproportionation between two activated chains.

A rise in temperature increases the rates of initiation and termination so that the rate of polymerisation is increased but the average chain-length of the polymer is reduced. The chain-length is also reduced by increasing the catalyst concentration, since this causes chain initiation to take place at many more points throughout the reaction mixture.

Condensation polymerisation involves chemical reactions between functional groups with the elimination of a small molecule, usually water. For example

$$x\underset{\text{hexamethylenediamine}}{NH_2(CH_2)_6NH_2} + x\underset{\text{adipic acid}}{COOH(CH_2)_4COOH} \rightarrow$$

$$\underset{\text{nylon 66}}{H[NH(CH_2)_6NH.CO(CH_2)_4CO]_xOH} + (2x-1)H_2O$$

If the monomers are bifunctional, as in the above example, then a linear polymer is formed. Terminating monofunctional groups will reduce the average degree of polymerisation. Polyfunctional monomers, such as glycerol and phthalic acid, are able to form branching points, which readily leads to irreversible network formation (see Chapter 9). Bakelite, a condensation product of phenol and formaldehyde, is an example of such a space-network polymer. Linear polymers are generally soluble in suitable solvents and are thermoplastic, i.e. they can be softened by heat without decomposition. In contrast, highly condensed network polymers are usually hard, almost completely insoluble and thermoset, i.e. they cannot be softened by heat without decomposition.

Dialysis

Conventional filter papers retain only particles with diameters in excess of at least $1\,\mu$m and are, therefore, permeable to colloidal particles.

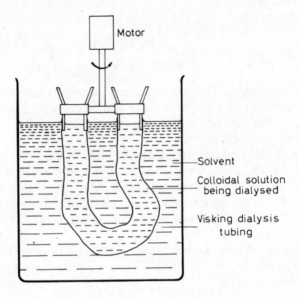

Figure 1.3. A simple dialysis set-up

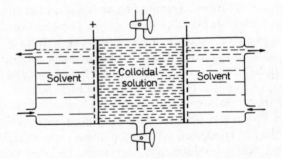

Figure 1.4. Electrodialysis

The use of membranes for separating particles of colloidal dimensions is termed *dialysis*.[19] The most commonly used membranes are prepared from regenerated cellulose products such as collodion (a partially evaporated solution of cellulose nitrate in alcohol plus ether), Cellophane and Visking. Membranes with various, approximately known, pore sizes can be obtained commercially (usually in the form of 'sausage skins' or 'thimbles'). However, particle size and pore size cannot be properly correlated, since the permeability of a membrane is also affected by factors such as electrical repulsion when the membrane and particles are of like charge, and particle adsorption on the filter which can lead to a blocking of the pores.

Dialysis is particularly useful for removing small dissolved molecules from colloidal solutions or dispersions, e.g. extraneous electrolyte such as KNO_3 from AgI sol. The process is hastened by stirring so as to maintain a high concentration gradient of diffusible molecules across the membrane and by renewing the outer liquid from time to time (Figure 1.3).

Ultrafiltration is the application of pressure or suction to force the solvent and small particles across a membrane while the larger particles are retained. The membrane is normally supported between fine wire screens or deposited in a highly porous support such as a sintered glass disc.

A further modification, *electrodialysis,* is illustrated in Figure 1.4. The applied potential between the metal screens supporting the membranes speeds up the migration of small ions to the membrane surface prior to their diffusion to the outer liquid. The accompanying concentration of charged colloidal particles at one side and, if they sediment significantly, at the bottom of the middle compartment is termed *electrodecantation*.

2
Kinetic Properties

THE MOTION OF PARTICLES IN LIQUID MEDIA

In this chapter the thermal motion and the motion under the influence of gravitational (or centrifugal) fields of colloidal molecules or particles dispersed in liquid (particularly aqueous) media will be considered. Thermal motion manifests itself on the microscopic scale in the form of Brownian motion, and on the macroscopic scale in the forms of diffusion and osmosis. Gravity (or a centrifugal field) provides the driving force in sedimentation. Among the most useful techniques for determining molecular or particle size and shape are those which involve the measurement of these simple properties.

The motion of colloidal particles in an electric field is treated separately in Chapter 7.

Before these kinetic properties are discussed in any detail, some general comments on the laws governing the motion of particles through liquids are appropriate.

Sedimentation rate

Consider the sedimentation of an uncharged particle of mass m and specific volume v in a liquid of density ρ. The driving (or sedimenting) force on the particle, which is independent of particle shape or solvation, is $m\,(1-v\rho)\,g$, where g is the local acceleration due to gravity (or a centrifugal field). The factor $(1-v\rho)$ allows for the buoyancy of the liquid. The liquid medium offers a resistance to the motion of the particle which increases with increasing velocity. Provided that the velocity is not too great, which is always the case for colloidal (and somewhat larger) particles, the resistance of the liquid is, to a first approximation, proportional to the velocity of the sedimenting particle. In a very short time, a terminal velocity, dx/dt, is attained, when the driving force on the particle and the resistance of the liquid are equal

$$m\,(1-v\rho)\,g = f\frac{dx}{dt} \qquad \ldots\,(2.1)$$

where f is the frictional coefficient for the particle in the given medium.

For spherical particles the frictional coefficient is given by Stokes' law

$$f = 6\pi\eta a \qquad \ldots\,(2.2)$$

where η is the viscosity of the medium and a the radius of the particle.

Therefore, if ρ_2 is the density of a spherical particle (in the dissolved or dispersed state, i.e. $\rho_2 = 1/v$), then

$$\tfrac{4}{3}\pi a^3\,(\rho_2-\rho)\,g = 6\pi\eta a\frac{dx}{dt}$$

or

$$\frac{dx}{dt} = \frac{2a^2\,(\rho_2-\rho)\,g}{9\eta} \qquad \ldots\,(2.3)$$

The derivation of Stokes' law assumes that:

1. The motion of the spherical particle is extremely slow.
2. The liquid medium extends an infinite distance from the particle, i.e. the solution or suspension is extremely dilute.
3. The liquid medium is continuous compared with the dimensions of the particle. This assumption is valid for the motion of colloidal particles, but not for that of small molecules or ions which are comparable in size with the molecules constituting the liquid medium.

For spherical colloidal particles undergoing sedimentation, diffusion or electrophoresis, deviations from Stokes' law usually amount to much less than 1 per cent and can be neglected.

Frictional ratios

The frictional coefficient of an asymmetric particle depends on its orientation. At low velocities such particles are in a state of random orientation through accidental disturbances, and the resistance of the liquid to their motion can be expressed in terms of a frictional coefficient averaged over all possible orientations. For particles of equal volume the frictional coefficient increases with increasing asymmetry. This is because, although the resistance of the liquid is reduced when the asymmetric particle is end-on to the direction of flow, it is increased to a greater extent with side-on orientations, so that on average there is an increase in resistance.

The frictional coefficient is also increased by particle solvation (hydration in aqueous systems).

A particle containing a given volume of dry material will have its smallest possible frictional coefficient, f_0, in a particular liquid when it is in the form of an unsolvated sphere. The *frictional ratio, f/f_0* (*i.e.* the ratio of the actual frictional coefficient to the frictional coefficient of the equivalent unsolvated sphere) is, therefore, a measure of a combination of asymmetry and solvation.

With application to dissolved proteins in mind, Oncley[20] has computed frictional ratios for ellipsoids of revolution of varying degrees of asymmetry and hydration. The resulting contour diagram (Figure 2.1) shows the combinations of axial ratio and hydration which are compatible with given frictional ratios. The separate contributions of asymmetry and hydration cannot be determined unless other relevant information is available.

BROWNIAN MOTION AND TRANSLATIONAL DIFFUSION

Brownian motion

A fundamental consequence of the kinetic theory is that, in the absence of external forces, all suspended particles, regardless of their size, have the same average translational kinetic energy. The average translational kinetic energy for any particle is $\frac{3}{2}kT$, or $\frac{1}{2}kT$ along a given axis, i.e. $\frac{1}{2}m\,(\mathrm{d}x/\mathrm{d}t)^2 = \frac{1}{2}kT$, etc.; in other words, the average particle velocity increases with decreasing particle mass.

The motion of individual particles is continually changing direction as a result of random collisions with the molecules of the suspending medium, other particles and the walls of the containing vessel. Each particle pursues a complicated and irregular zig-zag path. When the particles are large enough for observation this random motion is referred to as Brownian motion, after the botanist

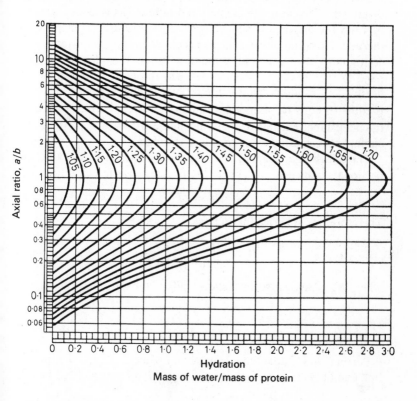

Figure 2.1. Values of axial ratio and hydration compatible with various frictional ratios (contour lines denote f/f_0 values) (*By courtesy of the authors*[20] *and Reinhold Publishing Corporation*)

who first observed this phenomenon with pollen grains suspended in water. The smaller the particles, the more evident is their Brownian motion.

Treating Brownian motion as a three-dimensional 'random walk', the mean Brownian displacement $\bar{x}$ of a particle from its original position along a given axis after a time t is given by Einstein's equation[21]

$$\bar{x} = \sqrt{2Dt} \qquad \ldots (2.4)$$

where D is the diffusion coefficient (see page 21).

The theory of random motion helps towards understanding the behaviour of linear high polymers in solution. The various segments of a flexible linear polymer molecule are subjected to independent thermal agitation, and so the molecule as a whole will take up a continually changing and somewhat random configuration (see page 6). The average distance between the ends of a completely flexible and random chain made up of n segments each of length l is equal to $l\sqrt{n}$ (cf. Einstein's equation above). This average end-to-end distance becomes $l\sqrt{2n}$ if an angle of $109°\,28'$ (the tetrahedral angle) between adjacent segments is specified.

The diffusion coefficient of a suspended material is related to the frictional coefficient of the particles by Einstein's law of diffusion

$$Df = kT \qquad \ldots (2.5)$$

Therefore, for spherical particles

$$D = \frac{kT}{6\pi\eta a} = \frac{RT}{6\pi\eta a N_A} \qquad \ldots (2.6)$$

where N_A is Avogadro's constant, and

$$\bar{x} = \sqrt{\frac{RTt}{3\pi\eta a N_A}} \qquad \ldots (2.7)$$

Perrin (1908) studied the Brownian displacement (and sedimentation equilibrium under gravity, see page 29) for fractionated mastic

Table 2.1 DIFFUSION COEFFICIENTS AND BROWNIAN DISPLACEMENTS CALCULATED FOR UNCHARGED SPHERES IN WATER AT $20°$C

Radius	$D_{20°C}/m^2\ s^{-1}$	$\bar{x}$ after 1 h
10^{-9} m (1 nm)	$2\cdot1 \times 10^{-10}$	$1\cdot23 \times 10^{-3}$ m (1·23 mm)
10^{-8} m (10 nm)	$2\cdot1 \times 10^{-11}$	$3\cdot90 \times 10^{-4}$ m (390 μm)
10^{-7} m (100 nm)	$2\cdot1 \times 10^{-12}$	$1\cdot23 \times 10^{-4}$ m (123 μm)
10^{-6} m (1 μm)	$2\cdot1 \times 10^{-13}$	$3\cdot90 \times 10^{-5}$ m (39 μm)

and gamboge suspensions of known particle size, and calculated values for Avogadro's constant varying between 5.5×10^{23} mol^{-1} and 8×10^{23} mol^{-1}. Subsequent experiments of this nature have yielded values of N_A closer to the accepted 6.02×10^{23} mol^{-1}; for example, Svedberg (1911) calculated $N_A = 6.09 \times 10^{23}$ mol^{-1} from observations on monodispersed gold sols of known particle size in the ultramicroscope. The correct determination of Avogadro's constant from observations on Brownian motion provides striking evidence in favour of the kinetic theory.

As a result of Brownian motion continual fluctuations of concentration take place on a molecular or small particle scale. For this reason the second law of thermodynamics is only valid on the macroscopic scale.

Translational diffusion

Diffusion is the tendency for molecules to migrate from a region of high concentration to a region of lower concentration and is a direct result of Brownian motion.

Fick's first law of diffusion (analogous with the equation for heat conduction) states that the mass of substance dm diffusing in the x direction in a time dt across an area A is proportional to the concentration gradient dc/dx at the plane in question

$$dm = -DA \frac{dc}{dx} dt \qquad \ldots . (2.8)$$

(The minus sign denotes that diffusion takes place in the direction of decreasing concentration.)

The rate of change of concentration at any given point is given by an exactly equivalent expression, Fick's second law

$$\frac{dc}{dt} = D \frac{d^2 c}{dx^2} \qquad \ldots . (2.9)$$

The proportionality factor D is called the diffusion coefficient. It is not strictly a constant since it is slightly concentration-dependent.

Equations (2.4) and (2.5) can be derived using Fick's first law equation (2.8), as follows:

1. *Brownian displacement equation* (2.4)—Consider a plane AB (Figure 2.2) passing through a dispersion and separating regions of

concentration c_1 and c_2, where $c_1 > c_2$. Let the average Brownian displacement of a given particle perpendicular to AB be $\bar{x}$ in time t. For each particle, this displacement has equal probability of being 'left to right' or 'right to left'. The net mass of particles displaced

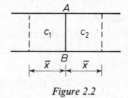

Figure 2.2

from left to right across unit area of AB in time t is, therefore, given by

$$m = \frac{(c_1 - c_2)\,\bar{x}}{2} = \frac{(c_1 - c_2)\,\bar{x}^2}{2\,\bar{x}}$$

If $\bar{x}$ is small,

$$\frac{c_1 - c_2}{\bar{x}} = -\frac{dc}{dx}$$

therefore,

$$m = -\frac{1}{2}\frac{dc}{dx}\bar{x}^2 \qquad \ldots (2.10)$$

From (2.8)

$$m = -D\frac{dc}{dx}t \qquad \ldots (2.11)$$

Therefore, combining Equations (2.10) and (2.11)

$$\bar{x} = \sqrt{2Dt} \qquad \ldots (2.4)$$

2. *Diffusion equation* (2.5)—The work done in moving a particle through a distance dx against a frictional resistance to motion $f\dfrac{dx}{dt}$ can be equated with the resulting change in chemical potential given by the expression

$$d\mu = kT\, d\ln c$$

i.e.

$$f\frac{dx}{dt}\, dx = kT\, d\ln c$$

therefore,

$$\frac{dx}{dt} = \frac{kT}{f}\frac{d\ln c}{dx} = \frac{kT}{fc}\frac{dc}{dx} \qquad \ldots (2.12)$$

Since $\qquad -\dfrac{\mathrm{d}m}{\mathrm{d}t} = Ac\dfrac{\mathrm{d}x}{\mathrm{d}t}$

then, combining this expression with Equation (2.8) gives

$$c\dfrac{\mathrm{d}x}{\mathrm{d}t} = D\dfrac{\mathrm{d}c}{\mathrm{d}x} \qquad\qquad \dots(2.13)$$

Therefore, combining Equations (2.12) and (2.13)

$$Df = kT \qquad\qquad \dots(2.5)$$

For a system containing spherical particles, $D = RT/6\pi\eta aN_A$, i.e. $D \propto 1/m^{1/3}$ where m is the particle mass. For systems containing asymmetric particles, D is correspondingly smaller (see Table 2.3). Since $D = kT/f$, the ratio D/D_0 (where D is the experimental diffusion coefficient and D_0 is the diffusion coefficient of a system containing the equivalent unsolvated spheres) is equal to the reciprocal of the frictional ratio f/f_0. Charge effects are discussed on page 30.

Measurement of diffusion coefficients[22]

1. *Free boundary methods*—To study free diffusion a sharp boundary must first be formed between the solution and the solvent

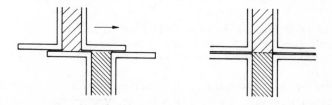

Figure 2.3. Formation of an initially sharp boundary between two miscible liquids

(or solution of lower concentration) in a suitable diffusion cell. One of the most satisfactory techniques for achieving this end is the shearing method illustrated in Figure 2.3. The boundary can be displaced away from the ground-glass flanges to facilitate optical observation, and can be sharpened even further by gently sucking away any mixed layers with an extremely fine capillary tube inserted from above.

As diffusion proceeds concentration and concentration gradient changes will take place as illustrated in Figure 2.4. To ensure that the broadening of the boundary is due to diffusion only, very accurate temperature control (to avoid convection currents) and freedom from mechanical vibration must be maintained. The avoidance of convection is a problem common to all kinetic methods of investigating colloidal systems.

Concentration changes are observed optically from time to time

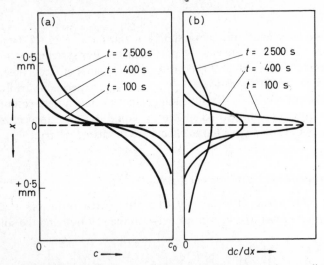

Figure 2.4. Distribution of (*a*) concentration and (*b*) concentration gradient at different times after the formation of a sharp boundary (calculated for $D = 2.5 \times 10^{-11} \text{ m}^2 \text{ s}^{-1}$)[4] (*By courtesy of Oxford University Press*)

either by light absorption (e.g. ultraviolet absorption for protein solutions), or, more usually, by schlieren or interference methods. These optical methods for examining concentration changes in liquid columns[4] (especially the schlieren technique in its various forms) are also employed in the ultracentrifuge and for studying moving-boundary electrophoresis. The schlieren method is based on the fact that, at a boundary between two transparent liquids of different refractive index, a beam of light perpendicular to the liquid column is refracted, thus casting a shadow which marks the region of changing refractive index. The optical system can be arranged so that the boundary is photographed in the form of a refractive index gradient peak. Since refractive index increments and concentration increments are normally proportional, the shape of the concentration gradient peak is also recorded directly.

Free diffusion columns are arranged to be sufficiently long for the initial concentrations at the extreme ends of the cell to remain un-altered during the course of the experiment. For a monodispersed system under these conditions the concentration gradient curves (Figure 2.4b) can be shown, by solving Fick's equations, to take the shape of Gaussian distribution curves represented by the expression

$$\frac{dc}{dx} = \frac{-c_0}{\sqrt{4\pi Dt}} \cdot \exp\left[-x^2/4Dt\right] \qquad \dots . (2.14)$$

from which D can be calculated. This Gaussian shape is not very sensitive to polydispersity, so that average diffusion coefficients can usually be calculated without too much difficulty.

2. *Porous plug method*—The two liquids are separated by a sintered glass disc with a pore size of $5-15\,\mu$m (the more concen-trated solution on top) and kept stirred. The liquid in the pores of

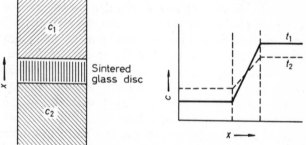

Figure 2.5. Porous plug method

the disc is effectively immobilized and freed from the influence of external disturbances, so that the transport of solute through the disc is due solely to diffusion. The extent of diffusion can be deter-mined by any analytical method.

By analogy with Fick's first law

$$\frac{dm}{dt} = \frac{-AD\,(c_1-c_2)}{l} \qquad \dots . (2.15)$$

where A is the cross-section of pores, and l is the effective length of pores.

The ratio A/l is determined by calibrating the apparatus with a substance of known diffusion coefficient.

This method has considerable advantages over the free boundary methods with regard to experimental procedure. Possible objections to the method are: (a) the calibration of the cell with material of different relative molecular mass and/or shape to the material under investigation is not necessarily valid; and (b) entrapment of air bubbles in the pores or adsorption of the diffusing molecules on the pore walls will invalidate the results.

THE ULTRACENTRIFUGE

There are a number of experimental techniques (for example, the Weigner tube and the Odén balance)[23, 24] which make use of sedimentation under gravity for fractionating or determining particle-size distributions in systems that contain relatively coarse suspended material such as soils and pigments. Sedimentation under gravity has a practical lower limit of ca. $1\mu m$. Smaller (colloidal) particles sediment so slowly under gravity that the effect is obliterated by the mixing tendencies of diffusion and convection.

By employing centrifugal forces instead of gravity, the application of sedimentation can be extended to the study of colloidal systems.[25-27] The driving force on a suspended molecule or particle then becomes $m(1-v\rho)\omega^2 x$, where ω is the angular velocity and x the distance of the particle from the axis of rotation.

An ultracentrifuge is a high-speed centrifuge equipped with a suitable optical system (usually schlieren) for recording sedimentation behaviour and with facilities for eliminating the disturbing

Table 2.2 SEDIMENTATION RATES UNDER GRAVITY FOR UNCHARGED SPHERES OF DENSITY 2 g cm^{-3} IN WATER AT 20°C, CALCULATED FROM STOKES' LAW

Radius	Sedimentation rate
10^{-9} m (1 nm)	$2 \cdot 2 \times 10^{-12}$ m s^{-1} (8 nm h^{-1})
10^{-8} m (10 nm)	$2 \cdot 2 \times 10^{-10}$ m s^{-1} (0·8 μm h^{-1})
10^{-7} m (100 nm)	$2 \cdot 2 \times 10^{-8}$ m s^{-1} (80 μm h^{-1})
10^{-6} m (1 μm)	$2 \cdot 2 \times 10^{-6}$ m s^{-1} (8 mm h^{-1})
10^{-5} m (10 μm)	$2 \cdot 2 \times 10^{-4}$ m s^{-1} (0·8 m h^{-1})

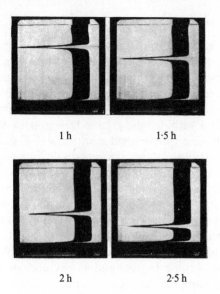

1 h   1·5 h

2 h   2·5 h

Figure 2.6. Sedimentation of a monodispersed sample of *Limulus* haemocyanin measured by the Philpot-Svensson schlieren method (18 000 rev min^{-1})[28] (By courtesy of the American Chemical Society)

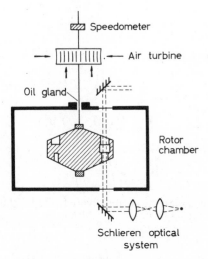

Figure 2.7. Essential features of an air-driven ultracentrifuge

effects of convection currents and vibration. The sample is contained in a sector-shaped cell mounted in a rotor (usually *ca.* 18 cm diameter); this spins in a thermostatted chamber containing hydrogen at a reduced pressure. Several mechanisms for driving the rotor have been investigated—Svedberg, who pioneered this field, employed an oil turbine; these have been superseded by simpler and less expensive air-driven and electrically-driven instruments.

The ultracentrifuge can be used in two distinct ways for investigating suspended colloidal material. In the velocity method a high centrifugal field (up to *ca.* 400 000 *g*) is applied and the displacement of the boundary set up by sedimentation of the colloidal molecules or particles is measured from time to time (Figure 2.6). In the equilibrium method the colloidal solution is subjected to a much lower centrifugal field, until sedimentation and diffusion (mixing) tendencies balance one another and an equilibrium distribution of particles throughout the sample is attained.

Sedimentation velocity

Equating the driving force on a macromolecule in a centrifugal field with the frictional resistance of the suspending medium

$$m (1-v\rho) \omega^2 x = f \frac{dx}{dt}$$

and since $Df = kT$

$$\frac{dx}{dt} = \frac{mD (1-v\rho) \omega^2 x}{kT}$$

$$= \frac{MD (1-v\rho) \omega^2 x}{RT}$$

or
$$M = \frac{RTs}{D (1-v\rho)} \qquad \dots (2.16)$$

where M is the molar mass (unsolvated), and s the sedimentation coefficient $= \dfrac{dx/dt}{\omega^2 x}$

Integrating
$$s = \frac{\ln x_2/x_1}{\omega^2 (t_2-t_1)} \qquad \dots (2.17)$$

where x_1 and x_2 are the distances of the boundary from the axis of rotation at times t_1 and t_2. Therefore

$$M = \frac{RT \ln x_2/x_1}{D\,(1-v\rho)\,(t_2-t_1)\,\omega^2} \qquad \ldots \ldots (2.18)$$

It is evident from the above expressions that the appropriate diffusion coefficient must also be measured in order that molecular or particle masses may be determined from sedimentation velocity data. In this respect, a separate experiment is required, since the diffusion coefficient cannot be determined accurately *in situ,* because there is a certain self-sharpening of the peak due to the sedimentation coefficient increasing with decreasing concentration.

Care must be taken to ensure that the system under investigation remains deflocculated. This applies to any technique for determining molecular or particle masses. s, D and v are corrected to a standard temperature, usually 20°C, and should be extrapolated to zero concentration.

With polydispersed systems either a broadening of the boundary (in addition to that caused by diffusion) or the formation of distinct peaks representing the various fractions is observed. Sedimentation does not provide an unequivocal method for establishing the homogeneity of a colloidal system. For example, a mixture of serum albumin and haemoglobin is homogeneous with respect to sedimentation velocity but the two proteins are easily distinguished from one another by electrophoresis.

Knowledge of M and v enables D_0, and hence the ratio D_0/D (the frictional ratio) to be calculated.

Sedimentation equilibrium

Consider the flow of molecules or particles across an area A in a colloidal solution where the concentration is c and the concentration gradient is dc/dx. The rate of flow is $cA\,(dx/dt)$ due to sedimentation, and, from Fick's first law, $-DA\,(dc/dx)$ due to diffusion. When sedimentation equilibrium is attained the net flow is zero, so that

$$c\frac{dx}{dt} = D\frac{dc}{dx}$$

and since

$$\frac{dx}{dt} = \frac{MD\,(1-v\rho)\,\omega^2 x}{RT}$$

then
$$\frac{dc}{c} = \frac{\omega^2 M (1-\nu\rho) \, x \, dx}{RT}$$

Integrating
$$M = \frac{2RT \ln c_2/c_1}{\omega^2 (1-\nu\rho) (x_2^2 - x_1^2)} \qquad \ldots . (2.19)$$

where c_1 and c_2 are the sedimentation equilibrium concentrations at distances x_1 and x_2 from the axis of rotation. Ideal behaviour has been assumed.

When a state of sedimentation–diffusion equilibrium has been reached, the molecular or particle mass can, therefore, be evaluated without a knowledge of the diffusion coefficient (and hence independently of shape and solvation) by determining relative concentrations at various distances from the axis of rotation. Molecules as small as sugars have been studied by this technique.

Polydispersity introduces complications and is reflected by a drift of M with x. Conversely, consistency of M with x is an indication of sample homogeneity with respect to M.

The disadvantage of the sedimentation equilibrium technique is that the establishment of equilibrium may take as long as several days, which is not only inconvenient generally but also accentuates the importance of avoiding convectional disturbances.

From a theoretical analysis of the intermediate stages during which the solute is being redistributed, Archibald[29] has developed a technique which involves measurements at intervals during the early stages of the sedimentation equilibrium experiment and so does not entail a long wait for equilibrium to be established. The ratio s/D can be calculated from the expression

$$\frac{dc/dx'}{c'x'} = \frac{\omega^2 s}{D} \qquad \ldots . (2.20)$$

where dc/dx' is the concentration gradient at the meniscus, c' the concentration at the meniscus, and x' the distance of the meniscus from the axis of rotation.

Charge effects

The treatment of sedimentation and diffusion is a little more complicated when the particles under consideration are charged. The smaller counter-ions (see Chapter 7) tend to sediment at a slower rate and lag behind the sedimentating colloidal particles. A potential is thus set up which tends to restore the original condition of overall electrical neutrality by accelerating the motion of the counter-ions and retarding the motion of the colloidal particles.

The reverse situation applies to diffusion. The smaller counter-ions tend to diffuse faster than the colloidal particles and drag the particles along with them and increase their rate of diffusion.

These effects can be overcome by employing swamping electrolyte concentrations. Any potentials which might develop are then readily dissipated by a very small displacement of a large number of counter-ions.

OSMOTIC PRESSURE

The measurement of a colligative property (i.e. lowering of vapour pressure, depression of freezing point, elevation of boiling point or osmotic pressure) is a standard procedure for determining the relative molecular mass of a dissolved substance. Of these, osmotic pressure is the only one with a practical value in the study of macromolecules. Consider, for example, a solution of 1 g of macromolecular material of relative molecular mass 50 000 dissolved in 100 cm^3 of water. Assuming ideal behaviour:

Depression of freezing point

$$= \frac{K_f c}{M} = \frac{1 \cdot 86 \text{ K kg mol}^{-1} \times 10^{-2}}{50\,000 \times 10^{-3} \text{ kg mol}^{-1}}$$

$$= 0 \cdot 0037 \text{ K}$$

Osmotic pressure at 20°C

$$= \frac{10 \text{ kg m}^{-3} \times 8 \cdot 314 \text{ J K}^{-1} \text{ mol}^{-1} \times 293 \text{ K}}{50\,000 \times 10^{-3} \text{ kg mol}^{-1}}$$

$$= 495 \text{ N m}^{-2} = 5 \text{ cm of water}$$

The above freezing point depression is far too small to be measured with sufficient accuracy by conventional methods and, even more important, it would be far too sensitive to small amounts of low relative molecular mass impurity; in fact, it would be doubled by the presence of just 1 mg of impurity of relative molecular mass 50.

Table 2.3 MOLECULAR DATA OF PROTEINS AND OTHER SUBSTANCES IN AQUEOUS SOLUTION

Name	$\dfrac{s_{20°C}}{10^{-13}\text{ s}}$	$\dfrac{D_{20°C}}{10^{-11}\text{ m}^2\text{ s}^{-1}}$	$\dfrac{v_{20°C}}{\text{cm}^3\text{ g}^{-1}}$	$M_r(s)$	$M_r(e)$	$M_r(\pi)$	f/f_0	Isoelectric point (pH)*
Urea		129				60		
Sucrose		36				342		
Ribonuclease	1·85	13·6	0·709	12700	13000		1·04	
Myoglobin	2·04	11·3	0·741	16900	17500	17000	1·11	7·0
Gliadin	2·1	6·7	0·724	27500	27000		1·6	
β-Lactoglobulin	3·1	7·3	0·751	41000	38000	35000	1·26	5·2
Ovalbumin	3·55	7·8	0·749	44000	40500	45000	1·16	4·55
Haemoglobin (horse)	4·48	6·3	0·749	68000	68000	67000	1·24	6·9
Serum albumin (horse)	4·46	6·1	0·748	70000	68000	73000	1·27	4·8
Serum globulin (horse)	7·1	4·0	0·745	167000	150000	175000	1·4	
Fibrinogen (bovine)	8·2	2·0	0·706	330000			2·3	5·2
Myosin	7·2	0·8	0·74	840000			4·0	5·4
Bushy stunt virus	132	1·15	0·739	10600000			1·27	4·1
Tobacco mosaic virus	174	0·3	0·727	59000000			2·9	

$M_r(s)$ = relative molecular mass from sedimentation velocity measurements
$M_r(e)$ = relative molecular mass from sedimentation equilibrium measurements
$M_r(\pi)$ = relative molecular mass from osmotic pressure measurements
*measured at 20°C in acetate or phosphate buffer at an ionic strength of 0·02 mol kg⁻¹

Not only does osmotic pressure provide a measurable effect, but also the effect of any low relative molecular mass material to which the membrane is permeable can be virtually eliminated.

The usefulness of osmotic pressure measurements is, nevertheless, limited to a relative molecular mass range of about 10^4 to 10^6. Below 10^4 permeability of the membrane to the molecules under consideration might prove to be troublesome, and above 10^6 the osmotic pressure will be too small to permit sufficiently accurate measurements.

Osmosis takes place when a solution and a solvent (or two solutions of different concentration) are separated from one another by a semipermeable membrane, i.e. a membrane which is permeable to the solvent but not to the solute. The tendency to equalize chemical potentials (and hence concentrations) on either side of the membrane results in a net diffusion of solvent across the membrane. The counter-pressure necessary to balance this osmotic flow is termed the *osmotic pressure*.

Osmosis can also take place in gels and constitutes an important swelling mechanism.

The osmotic pressure Π of a solution is described in general terms by the so-called virial equation

$$\Pi = cRT \left(\frac{1}{M} + B_2\, c + B_3\, c^2 + \ldots\right) \quad \ldots\ (2.21)$$

where c is the concentration of the solution, M is the molar mass of the solute and B_2, B_3, etc. are constants.

Therefore, $$M = RT / \lim_{c \to 0} \Pi/c \quad \ldots\ (2.22)$$

Deviations from ideal behaviour are relatively small for solutions of compact macromolecules such as proteins but can be quite appreciable for solutions of linear polymers. Such deviations have been treated thermodynamically,[30, 31] mainly in terms of the entropy change on mixing, which is considerably greater (especially for linear polymers dissolved in good solvents) than the ideal entropy change on mixing for a system obeying Raoult's law. This leads to solvent activities which are smaller than ideal, i.e. an apparent increase in the concentration, and an actual increase in the osmotic pressure of the polymer solution.

The resulting relative molecular mass refers to the composition of the solute with respect to solvation, etc., which was used in establishing the concentration of the solution. For polydispersed systems a number average is measured.

Measurement of osmotic pressure

A great deal of work has been devoted to the preparation of suitable semipermeable membranes and to perfecting sensitive methods for measuring osmotic pressure.[32-34] A simple set-up for measuring the osmotic pressure of an aqueous solution is shown in Figure 2.9.

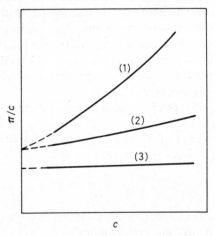

Figure 2.8. Dependence of reduced osmotic pressure on concentration
(1) A linear high polymer in a good solvent
(2) The same polymer in a poor solvent
(3) A globular protein in aqueous solution

Certain practical difficulties arise if the solution is simply allowed to rise and seek its own equilibrium level:

1. If the liquid rises up a narrow tube and the total volume of solution is large then the liquid level may be too sensitive to temperature fluctuations during the course of the experiment to permit reliable measurements. In addition, a capillary rise correction must be made under these conditions. To overcome the difficulty associated with a sticking meniscus, which is often encountered when studying aqueous solutions, a liquid of low surface tension and

good wetting properties, such as toluene or petrol ether, is used in the capillary.

2. If the liquid rises up a wide tube and the volume of solution is small, significant changes in concentration will take place and the establishment of equilibrium will be extremely slow.

A frequently adopted procedure for overcoming these difficulties is to set the liquid level in turn at slightly above and slightly below the anticipated equilibrium level and then plot its position as a function of time. If the estimation of the final equilibrium level has been reasonably accurate, then the two curves will be almost symmetrical, and by plotting the half-sum as a function of time the equilibrium level can be reliably estimated after relatively short times (Figure 2.10). This procedure permits the use of a moderately wide capillary tube; however, good thermostatting and a firmly supported membrane are still required.

In the Fuoss-Mead osmometer[35] the membrane is firmly clamped (and also acts as a gasket) between two carefully machined stainless steel blocks provided with channels into which small volumes of

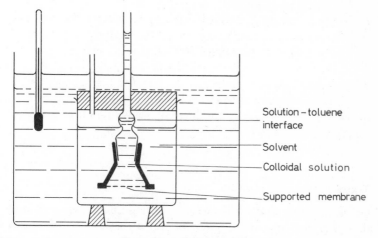

Solution – toluene interface

Solvent

Colloidal solution

Supported membrane

Figure 2.9. A simple osmometer

solvent and solution are introduced (Figure 2.11). Owing to the high ratio of membrane surface area to solution volume (*ca.* 75 cm^2 : 15 cm^3) the approach to equilibrium is rapid. With fast membranes, and using the half-sum method, measurements in well under 1 hour are possible. Over such a short time the membrane can be slightly permeable to the macromolecules without introducing serious errors.

A different principle of measurement is employed in the osmotic balance (Figure 2.12). The compartment of the osmometer containing the solution is suspended from the arm of a balance, and the

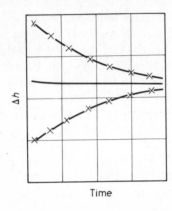

Figure 2.10. Estimation of osmotic pressure of the half-sum method

osmotic pressure determined by weighing. With careful experimentation, especially with respect to temperature stability, a precision of better than 0·01 mm of solution is attainable, which makes this method particularly useful for determining small osmotic pressures. Because of the length of time required for the establishment of equilibrium, it is usual to employ a dynamic technique (e.g. the half-sum method, or measuring the flow rate for different heads of solution and interpolating to zero flow).

The Donnan membrane equilibrium

Certain complications arise when solutions containing both non-diffusible and (inevitably) diffusible ionic species are considered. Gibbs predicted and later Donnan demonstrated that when the non-diffusible ions are located on one side of a semipermeable membrane the distribution of the diffusible ions is unequal when equilibrium is attained, being greater on the side of the membrane containing the non-diffusible ions. This distribution can be calculated thermodynamically, although a simpler kinetic treatment will suffice.

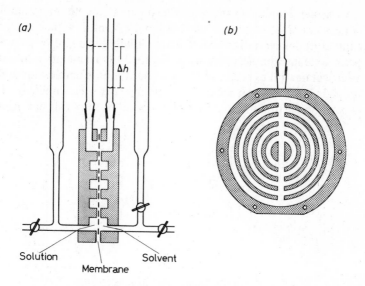

Figure 2.11. Schematic representation of the Fuoss-Mead osmometer: (*a*) vertical cross-section; (*b*) inner surface of each half-cell

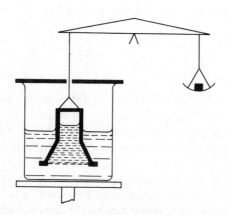

Figure 2.12. The osmotic balance

Consider a simple example in which equal volumes of solutions of the sodium salt of a protein and of sodium chloride with respective equivalent concentrations a and b are initially separated by a semipermeable membrane, as shown in Figure 2.13. To maintain overall electrical neutrality Na^+ and Cl^- ions must diffuse across the membrane in pairs. The rate of diffusion in any particular direction will depend on the probability of a Na^+ and a Cl^- ion arriving at a given

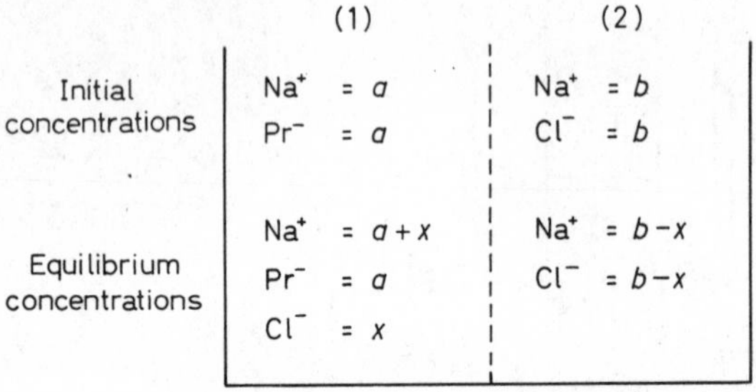

	(1)	(2)
Initial concentrations	$Na^+ = a$ $Pr^- = a$	$Na^+ = b$ $Cl^- = b$
Equilibrium concentrations	$Na^+ = a+x$ $Pr^- = a$ $Cl^- = x$	$Na^+ = b-x$ $Cl^- = b-x$

Figure 2.13. The Donnan membrane equilibrium

point on the membrane surface simultaneously. This probability is proportional to the product of the Na^+ and Cl^- ion concentrations (strictly, activities) so that

$$\text{Rate of diffusion from (1) to (2)} = k\,(a+x)\,x$$

$$\text{Rate of diffusion from (2) to (1)} = k\,(b-x)^2$$

At equilibrium, these rates of diffusion are equal

i.e. $$(a+x)\,x = (b-x)^2$$

or $$x = \frac{b^2}{a+2b}$$

At equilibrium the concentrations of diffusible ions in compartments (1) and (2) are $(a+2x)$ and $2\,(b-x)$ respectively, so that the excess concentration in compartment (1) is $(a-2b+4x)$. Substituting for x, this excess diffusible ion concentration works out to be $a^2/(a+2b)$.

Clearly, the results of osmotic pressure measurements on solutions of charged colloidal particles, such as proteins, will be vitiated unless precautions are taken either to eliminate or to correct for this Donnan effect. Working at the isoelectric pH of the protein will eliminate the Donnan effect but will probably introduce new errors due to coagulation of the protein. Working with a moderately large salt concentration and a small protein concentration will make the ratio $a^2/(a+2b)$ small and allow the Donnan effect to be virtually eliminated.

ROTARY BROWNIAN MOTION

In addition to translational Brownian motion, suspended molecules or particles undergo random rotational motion about their axes, so that, in the absence of aligning forces, they are in a state of random orientation. Rotary diffusion coefficients can be defined (ellipsoids of revolution have two such coefficients representing rotation about

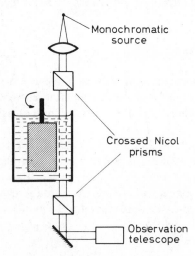

Figure 2.14. Apparatus for observing streaming birefringence

each principal axis) which depend on the size and shape of the molecules or particles in question.[4]

Under the influence of an orientating force partial alignment of asymmetric particles takes place, which represents a balance between the aligning force on the particles and their rotary diffusion. The system will therefore become anisotropic. It is possible to draw conclusions concerning particle dimensions by studying various changes in physical properties brought about by such particle alignment.

Streaming birefringence

The sample is subjected to a strong velocity gradient—for example, in a concentric cylinder viscometer (Figure 2.14)—and the resulting molecular or particle alignment causes the previously isotropic solution to become doubly refracting (birefringent). The magnitude of the birefringence is related in an elaborate theory to the rotational diffusion coefficient, and hence to the molecular or particle dimensions.

In certain cases, e.g. ferric hydroxide sol, birefringence can be produced by the aligning action of electrical or magnetic fields.

Dielectric dispersion

When a solution containing dipolar molecules is placed between electrodes and subjected to an alternating current, the molecules tend to rotate in phase with the current, thus increasing the dielectric constant of the solution. As the frequency is increased it becomes more difficult for the dipolar molecules to overcome the viscous resistance of the medium rapidly enough to remain in phase, and the dielectric constant drops in a more or less stepwise fashion. Each characteristic frequency where there is a notable change of dielectric constant is related to the time taken for the molecule to rotate about a particular axis, and hence to the appropriate rotary diffusion coefficient.

3
Optical Properties

LIGHT SCATTERING

When a beam of light is directed at a colloidal solution or dispersion some of the light may be absorbed (colour is produced when light of certain wavelengths is selectively absorbed), some is scattered, whilst the remainder is transmitted undisturbed through the sample.

The Tyndall effect—turbidity

All materials are capable of scattering light (Tyndall effect) to some extent. The noticeable turbidity associated with many colloidal dispersions is a consequence of intense light scattering. A beam of sunlight is often visible from the side because of light scattered by dust particles. Solutions of certain macromolecular materials may appear to be clear, but in fact they are slightly turbid because of weak light scattering. Only a perfectly homogeneous system would not scatter light; therefore, even pure liquids and dust-free gases are very slightly turbid.

The turbidity of a material is defined by the expression

$$I_t/I_0 = \exp[-\tau l] \qquad \qquad \dots (3.1)$$

where I_0 is the intensity of the incident light beam, I_t that of the transmitted light beam, l the length of the sample, and τ the turbidity.

Measurement of scattered light

As we shall see, the intensity, polarisation and angular distribution of the light scattered from a colloidal dispersion depend on the size and shape of the scattering particles, the interactions between them, and the difference between the refractive indices of the particles and the medium. Light scattering measurements are, therefore, of great value for estimating particle size, shape and interactions, and have found particular application in the study of dissolved macromolecular material.[36]

The intensity of the light scattered by colloidal solutions or suspensions of low turbidity is measured directly. A detecting photocell is generally mounted on a rotating arm to permit measurement of the light scattered at several angles, and fitted with a polaroid for observing the polarization of the scattered light (see Figure 3.1). Weakening of the scattered beam itself as it passes through the

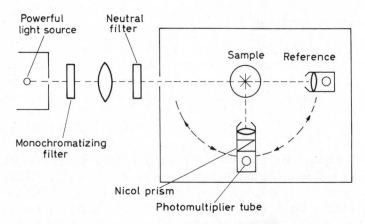

Figure 3.1. Measurement of scattered light

slightly turbid sample can be neglected, and its intensity can be compared with that of the transmitted beam.

Although simple in principle, light-scattering measurements present a number of experimental difficulties, the most notable being the necessity to free the sample from impurities such as dust, the relatively large particles of which would scatter light strongly and introduce serious errors.

Scattering by small particles

Rayleigh (1871) laid the foundation of light-scattering theory by applying the electromagnetic theory of light to the scattering by small, non-absorbing (insulating), spherical particles in a gaseous medium. When an electromagnetic wave of intensity I_0 and wavelength λ falls on a small ($<$ *ca.* $\lambda/20$) particle of polarisability α,

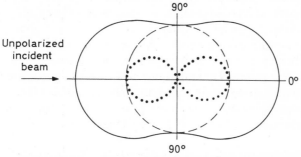

Figure 3.2. Radiation envelope for light scattered from small particles. Distances from the origin to the dotted, dashed and smooth lines represent the relative intensities of the horizontally polarised component, vertically polarised component and total scattered light respectively

oscillating dipoles are induced in the particle. The particle then serves as a secondary source for the emission of scattered radiation of the same wavelength as the incident light. For an unpolarised incident beam, the intensity I_θ at a distance r from the particle of the light scattered at an angle θ to the incident beam is given by the expression

$$\frac{I_\theta r^2}{I_0} = \frac{8\pi^4\alpha^2}{\lambda^4}(1+\cos^2\theta) = R_\theta(1+\cos^2\theta) \quad \ldots. (3.2)$$

The quantity $R_\theta(1+\cos^2\theta)$ is called the Rayleigh ratio. The unity term in $(1+\cos^2\theta)$ refers to the vertically polarised component of the scattered light, and the $\cos^2\theta$ term to the horizontally polarised component.

Since the scattering intensity is proportional to $1/\lambda^4$, then blue light ($\lambda \sim 450$ nm) is scattered much more than red light ($\lambda \sim 650$ nm). With incident white light, a scattering material will, therefore, tend to appear to be blue when viewed at right-angles to the incident beam, and red when viewed from end-on. This phenomenon is evident in the blue colour of the sky, tobacco smoke, diluted milk, etc., and the yellowish-red of the rising and setting sun.

Interparticle interference

If the scattering sources in a system are close together and regularly spaced, as in a crystalline material, then there will be regular phase relationships (coherent scattering) and, therefore, almost total destructive interference between the scattered light waves, i.e. the intensity of the resulting scattered light will be almost zero. When the scattering sources are randomly arranged, which is virtually the case for gases, pure liquids and dilute solutions or dispersions, there are no definite phase relationships (incoherent scattering) and destructive interference between the scattered light waves is incomplete.

For a system of independent scatterers (point sources of scattered. light distributed completely at random), the emitted light waves have an equal probability of reinforcing or destructively interfering with one another. The amplitudes of the scattered waves add and subtract in a random fashion, with the result that (by analogy with Brownian displacement, page 20) the amplitude of the total scattered light is proportional to the square root of the number of scattering particles. Since the intensity of a light wave is proportional to the square of its amplitude, the total intensity of scattered light is proportional to the number of particles.

Relative molecular masses from light scattering measurements

If the dimensions of a scattering particle are all less than *ca.* $\lambda/20$, then the scattered light waves emanating from the various parts of the particle cannot be more than *ca.* $\lambda/10$ out of phase, and so their amplitudes are practically additive. The total amplitude of the light scattered from such a particle is, therefore, proportional to the number of individual scatterers in the particle, i.e. to its volume, and hence its mass; and the total intensity of scattered light is proportional to the square of the particle mass. Consequently, for a random dispersion containing n particles of mass m, the total amount of light scattered is proportional to nm^2; and as nm is proportional to the concentration c of the dispersed phase

$$\text{total light scattered} \propto cm$$

An alternative (but equivalent) approach is the so-called fluctuation theory in which light scattering is treated as a consequence of random non-uniformities of concentration, and hence refractive index, arising from random molecular movement (see page 21). Using this approach, the above relationship can be written in the quantitative form derived by Debye[37] for dilute macromolecular solutions:

$$\frac{Hc}{\tau} = \frac{1}{M} + 2Bc$$

i.e.

$$\frac{Hc}{\tau}\bigg|_{\lim c \to 0} = \frac{1}{M} \qquad \ldots\ldots (3.3)$$

where τ is the turbidity of the solution, M is the molar mass of the solute and H is a constant given by

$$H = \frac{32\pi^3 n_o^2}{3N_A \lambda_o^4} \left(\frac{dn}{dc}\right)^2 \qquad \ldots\ldots (3.4)$$

where n_o is the refractive index of the solvent, n that of the solution and λ_o is the wavelength *in vacuo* (i.e. $\lambda_o = n\lambda$, where λ is the wavelength of the light in the solution). τ is calculated from the intensity of light scattered at a known angle (usually 90° or 0°). Summation of the products, $Rd\omega$, over the solid angle 4π leads to the relationship

$$\tau = \frac{16\pi}{3} R_{90°} \qquad \ldots\ldots (3.5)$$

where R [defined in Equation (3.2)] now refers to primary scattering from unit volume of solution.

Therefore,

$$\frac{Kc}{R_{90°}}\bigg|_{\lim c \to 0} = \frac{1}{M} \qquad \ldots\ldots (3.6)$$

where

$$K = \frac{2\pi^2 n_o^2}{N_A \lambda_o^4} \left(\frac{dn}{dc}\right)^2 \qquad \ldots\ldots (3.7)$$

dn/dc is measured accurately using a differential refractometer.

In contrast to osmotic pressure, light-scattering measurements become easier as the particle size increases. For spherical particles the upper limit of applicability of the Debye equation is a particle diameter of *ca.* $\lambda/20$ (i.e. 20 nm to 25 nm for $\lambda_o \sim 600$ nm, or $\lambda_{water} \sim 450$ nm; or a relative molecular mass of the order of 10^7). For asymmetric particles this upper limit is lower. However, by modification of the theory, much larger particles can also be studied by light-scattering methods. For polydispersed systems a mass-average relative molecular mass is given.

Large particles

The theory of light scattering is more complicated when one or more of the particle dimensions exceeds *ca.* $\lambda/20$. Such particles cannot be considered as point sources of scattered light, and destructive interference between scattered light waves originating from different locations on the same particle must be taken into account. This intraparticle destructive interference is zero for light scattered in a forward direction ($\theta = 0°$). Extinction will

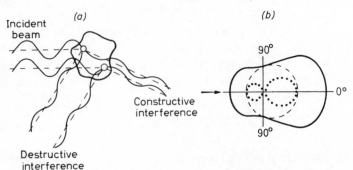

Figure 3.3. (*a*) Scattering from a relatively large particle. (*b*) Radiation envelope for light scattered from a spherical particle ($x = 0.8$, $m = 1.25$) – see text and *Figure 3.2* for explanation

take place between waves scattered backwards from the front and rear of a spherical particle of diameter $\lambda/4$ (i.e. the total path difference is $\lambda/2$). The radiation envelope for such a particle will, therefore, be unsymmetrical, more light being scattered forwards than backwards.

When particles of significantly different refractive index from that of the suspending medium contain a dimension greater than *ca.* $\lambda/4$, extinction at intermediate angles is possible and maxima and minima of scattering can be observed at different angles. The location of such maxima and minima will depend on the wavelength, so that with incident white light it is possible, with a

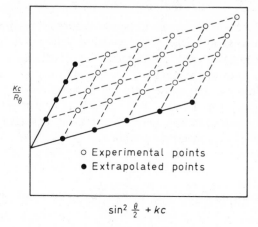

$$\sin^2 \tfrac{\theta}{2} + kc$$

Figure 3.4. A Zimm plot

suitable monodispersed system, to observe spectral colour sequences (known as 'higher-order Tyndall spectra').

Mie (1908) elaborated a general quantitative theory for light scattering by spherical particles. The intensity of the light scattered at various angles is related to *m*, the ratio of the refractive index of the particles to that of the medium, and the parameter[38] $x = 2\pi r/\lambda$. Mie's theory has been extended by Gans to include certain non-spherical shapes. The main features of Mie's theory have been verified by LaMer *et al*[39] from measurements of the angular variation of the light scattered by monodispersed sulphur sols containing particles of radius 300 nm to 600 nm.

Since the light scattered forwards (0°) suffers no intraparticle interference, its intensity is proportional to the square of the particle mass. By measuring the light scattered by a colloidal solution or dispersion as a function of both angle and concentration and extrapolating to zero angle and zero concentration, the size of relatively large particles can be calculated from the Debye equation. This extrapolation (Zimm plot,[40] Figure 3.4) is achieved by plotting Kc/R_θ against $\sin^2 \frac{\theta}{2} + kc$, where k is an arbitrary constant selected so as to give convenient spacing between the points on the graph.

$$\frac{Kc}{R_\theta}_{\substack{\lim c \to 0 \\ \theta \to 0}} = \frac{1}{M} \qquad \qquad \dots (3.8)$$

Particles which are too small to show a series of maxima and minima in the angular variation of scattered light are frequently studied by measuring the dissymmetry of scattering (usually defined as the ratio of the light scattered at 45° to that scattered at 135°). The dissymmetry of scattering is a measure of the extent of the particles compared with λ. If the molecular or particle size is known it can be related to the axial ratio of rod-like particles, or the coiling of flexible linear macromolecules.

ELECTRON MICROSCOPY AND DARK-FIELD MICROSCOPY

Resolving power

Colloidal particles are generally too small to permit direct microscopic observation. The resolving power of an optical microscope (i.e. the smallest distance by which two objects may be separated and yet remain distinguishable from one another) is limited mainly by the wavelength λ of the light used for illumination. The limit of resolution δ is given by the expression

$$\delta = \lambda/2n \sin \alpha \qquad \qquad \dots (3.9)$$

where α is the angular aperture (half the angle subtended at the object by the objective lens), n the refractive index of the medium between the object and the objective lens, and $n \sin \alpha$ the numerical aperture of the objective lens for a given immersion medium.

The numerical aperture of an optical microscope is generally less than unity. With oil-immersion objectives numerical apertures up to about 1·6 are attainable, so that, for light of wavelength 600 nm, this would permit a resolution limit of about 200 nm (0·2μm). Since the human eye can readily distinguish objects some 0·2 mm (200 μm) apart, there is little advantage in using an optical microscope, however well constructed, which magnifies greater than about 1 000 times. Further magnification increases the size but not the definition of the image.

In addition to the question of resolving power, the visibility of an object may be limited due to lack of optical contrast between the object and its surrounding background.

Two techniques for overcoming these difficulties are of particular value in the study of colloidal systems. They are electron microscopy, in which the limit of resolution is greatly extended, and dark-field microscopy, in which the minimum observable contrast is greatly reduced.

The electron microscope[41—44]

To increase the resolving power of a microscope so that matter of colloidal (and smaller) dimensions may be observed directly, the wavelength of the radiation used must be considerably reduced

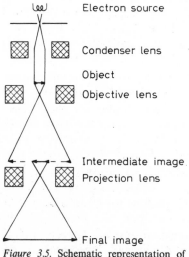

Figure 3.5. Schematic representation of the electron microscope

below that of visible light. Electron beams can be produced with wavelengths of the order of 0·01 nm and focused by electric or magnetic fields, which act as the equivalent of lenses. The resolution of an electron microscope is limited not so much by wavelength, but by the technical difficulties of stabilising high-tension supplies and correcting lens aberrations. Only lenses with a numerical aperture of less than 0·01 are usable at present. The best resolution which has been attained so far is about 0·5 nm, which is not much greater than atomic dimensions.

The use of the electron microscope for studying colloidal systems is limited by the fact that electrons can only travel unhindered in high vacuum, so that any system having a significant vapour pressure must be thoroughly dried before it can be observed. Such pretreatment may result in a misrepresentation of the sample under consideration. Instability of the sample to electron beams could also result in misrepresentation.

A small amount of the material under investigation is deposited on an electron-transparent plastic or carbon film (10 nm to 20 nm thick) supported on a fine copper mesh grid. The sample scatters electrons out of the field of view, and the final image can be made visible on a fluorescent screen. The amount of scattering depends on the thickness and on the atomic number of the atoms forming the specimen, so that organic materials are relatively electron-transparent and show little contrast against the background support, whereas materials containing heavy metal atoms make ideal specimens.

To enhance contrast and obtain three-dimensional effects the technique of shadow-casting is generally employed. A heavy metal, such as gold, is evaporated in vacuum and at a known angle on to the specimen, thus giving a side illumination effect (see Figure 3.6). From the angle of shadowing and the length of the shadows a three-dimensional picture of the specimen can be built up. An even better picture can be obtained by lightly shadowing the sample in two directions at right-angles.

A most useful technique for examining surface structure is that of replication. One method is to deposit the sample on a freshly cleaved mica surface on to which carbon (and, if desired, a heavy metal) is vacuum-evaporated. The resulting thin film, with the specimen particles still embedded, is floated off the mica on to a water surface. The particles are dissolved out with a suitable solvent and the resulting replica is mounted on a copper grid.

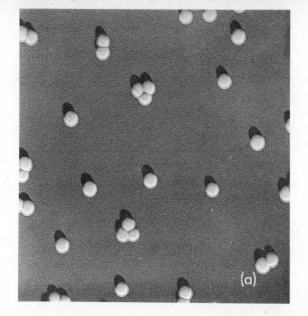

Figure 3.6. Electron micrographs. (*a*) Shadowed polystyrene latex particles (× 50000). (*b*) Shadowed silver chloride particles (× 15000)

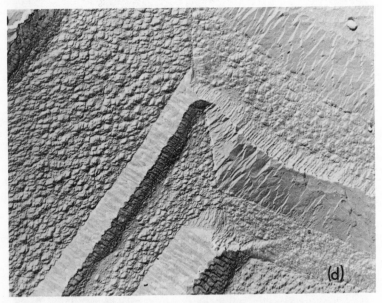

Figure 3.6. Electron micrographs. (*c*) Platelets of nordstrandite (aluminium hydroxide) (× 5000). (*d*) Replica of an etched copper surface (× 5600)

Dark-field microscopy—the ultramicroscope

Dark-field illumination is a particularly useful technique for detecting the presence, counting and investigating the motion of suspended colloidal particles. It is obtained by arranging the illumination system of an ordinary microscope so that light does

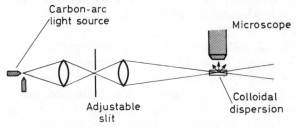

Figure 3.7. Principle of the slit ultramicroscope

not enter the objective unless scattered by the sample under investigation.

If the particles in a colloidal dispersion have a refractive index sufficiently different from that of the suspending medium, and an intense illuminating beam is used, sufficient light is deflected into

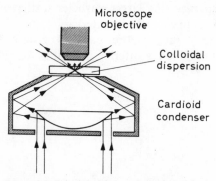

Figure 3.8. Principle of the cardioid darkfield condenser

the objective for the particles to be observed as bright specks against a dark background. Lyophobic particles as small as 5 nm to 10 nm can be made indirectly visible in this way. Owing to solvation, the refractive index of lyophilic particles, such as dissolved macromolecules, is little different from that of the suspending medium, and they scatter insufficient light for detection by dark field methods.

The two principal techniques of dark-field illumination are the slit and cardioid methods. In the slit ultramicroscope of Siedentopf and Zsigmondy (1903) the sample is illuminated from the side by an intense narrow beam of light from a carbon-arc source (Figure 3.7). The cardioid condenser (a standard microscope accessory) is an optical device for producing a hollow cone of illuminating light; the sample is located at the apex of the cone where the light intensity is high (Figure 3.8).

Dark-field methods do not help to improve the resolving power of a microscope. A small scattering particle is seen indirectly as a weak blur. Two particles must be separated by the resolution distance δ to be separately visible. Dark-field microscopy is, nevertheless, an extremely useful technique for studying colloidal dispersions and obtaining information concerning:

1. Brownian motion.
2. Sedimentation equilibrium.
3. Electrophoretic mobility.
4. The progress of flocculation.
5. Number average particle size (from counting experiments and a knowledge of the concentration of dispersed phase).
6. Polydispersity (the larger particles scatter more light and, therefore, appear to be brighter).
7. Asymmetry (asymmetric particles give a flashing effect owing to different scattering intensities for different orientations).

4

Liquid–Gas and Liquid–Liquid Interfaces

SURFACE AND INTERFACIAL TENSIONS

It is well known that short-range van der Waals forces of attraction exist between molecules (see page 169), and are responsible for the existence of the liquid state. The phenomena of surface and interfacial tension are readily explained in terms of these forces. The molecules which are located within the bulk of a liquid are, on average, subjected to equal forces of attraction in all directions, whereas those located, for example, at a liquid–air interface experience unbalanced attractive forces resulting in a net inward pull (Figure 4.1). As many molecules as possible will leave the liquid surface for the interior of the liquid; the surface will, therefore, tend to contract spontaneously. For this reason droplets of liquid and bubbles of gas tend to attain a spherical shape.

Surface tension (and the more fundamental quantity, *surface free energy*) fulfil an outstanding role in the physical chemistry of surfaces. The surface tension γ_0 of a liquid is often defined as the force acting at right-angles to any line of unit length on the liquid surface. However, this definition (although appropriate in the case of liquid films, such as in foams) is somewhat misleading, since there is no elastic skin or tangential force as such at the surface of a pure liquid. It is more satisfactory to define surface tension and surface free energy as the work required to increase the area of a surface isothermally and reversibly by unit amount.

The same considerations apply to the interface between two immiscible liquids. Again there is an imbalance of intermolecular

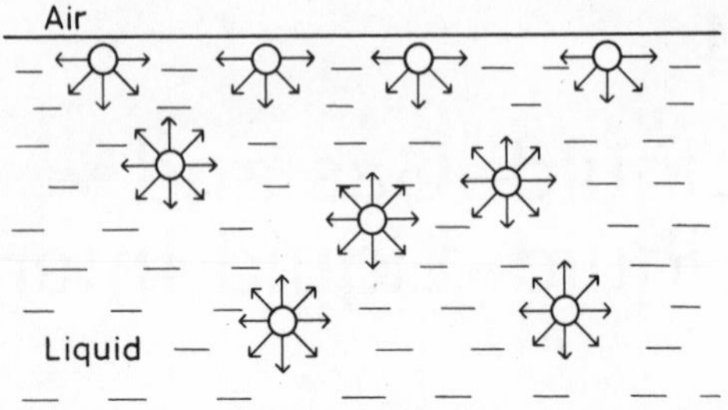

Figure 4.1. Attractive forces between molecules at the surface and in the interior of a liquid

forces but of a lesser magnitude. Interfacial tensions usually lie between the individual surface tensions of the two liquids in question.

The above picture implies a static state of affairs. However, it must be appreciated that an apparently quiescent liquid surface is actually in a state of great turbulence on the molecular scale as a result of two-way traffic between the bulk of the liquid and the surface, and between the surface and the vapour phase.[45] The average lifetime of a molecule at the surface of a liquid is *ca.* 10^{-6} seconds.

Table 4.1   SURFACE TENSIONS AND INTERFACIAL TENSIONS AGAINST WATER FOR LIQUIDS AT 20°C (IN mN m^{-1})

Liquid	γ_0	γ_i	Liquid	γ_0	γ_i
Water	72·75	—	Ethanol	22·3	—
Benzene	28·88	35·0	n-Octanol	27·5	8·5
Acetic acid	27·6	—	n-Hexane	18·4	51·1
Acetone	23·7	—	n-Octane	21·8	50·8
CCl$_4$	26·8	45·1	Mercury	485	375

Phenomena at curved interfaces

As a consequence of surface tension there is a balancing pressure difference across any curved surface, the pressure being greater on the concave side. For a curved surface with principal radii of curvature r_1 and r_2 this pressure difference is given by the Young–Laplace equation, $\Delta p = \gamma \, (1/r_1 + 1/r_2)$, which reduces to $\Delta p = 2\gamma/r$ for a spherical surface.

The vapour pressure over a small droplet (where there is a high surface/volume ratio) is higher than that over the corresponding flat surface. The transfer of liquid from a plane surface to a droplet requires the expenditure of energy since the area, and hence the surface free energy, of the droplet will increase.

If the radius of a droplet increases from r to $r+dr$, then the surface area will increase from $4\pi r^2$ to $4\pi \, (r+dr)^2$, i.e. by $8\pi r \, dr$, and the increase in surface free energy will be $8\pi \gamma r \, dr$. If this process involves the transfer of dn moles of liquid from the plane surface with a vapour pressure p_0 to the droplet with a vapour pressure p_r, the free energy increase is also equal to $dnRT \ln p_r/p_0$, assuming ideal gaseous behaviour. Equating these free energy increases

$$dnRT \ln p_r/p_0 = 8\pi \gamma r \, dr$$

and since

$$dn = 4\pi r^2 \, dr \, \rho/M$$

then

$$RT \ln p_r/p_0 = \frac{2\gamma M}{\rho r} = \frac{2\gamma V_m}{r} \qquad \dots\, (4.1)$$

where ρ is the density of the liquid, V_m the molar volume of the liquid, and M the molar mass. For example, for water droplets (assuming γ to be constant)

$$r = 10^{-7} \text{ m} \qquad p_r/p_0 \sim 1\cdot01$$
$$10^{-8} \text{ m} \qquad\qquad 1\cdot1$$
$$10^{-9} \text{ m} \qquad\qquad 3\cdot0$$

This expression, known as the Kelvin equation, has been verified experimentally. It can also be applied to a concave capillary meniscus; in this case the curvature is negative and a vapour pressure lowering is predicted.

The effect of curvature on vapour pressure (and, similarly, on solubility) provides a ready explanation for the ability of vapours (and solutions) to supersaturate. If condensation has to take place via droplets containing only a few molecules, the high vapour pressures involved will present an energy barrier to the process, whereas in the presence of foreign matter this barrier can be by-passed.

Variation of surface tension with temperature

The surface tension of most liquids decreases with increasing temperature in a nearly linear fashion (some metal melts being exceptional in this respect) and becomes very small in the region of the critical temperature when the intermolecular cohesive forces approach zero. A number of empirical equations have been suggested which relate surface tension and temperature, one of the most satisfactory being that of Ramsay and Shields

$$\gamma \left(\frac{Mx}{\rho} \right)^{2/3} = k\,(T_c - T - 6) \qquad \dots (4.2)$$

where M is the molar mass of the liquid, ρ the density of the liquid, x the degree of association of the liquid, T_c the critical temperature, and k a constant.

Measurement of surface and interfacial tensions

1. *Capillary rise method*—This is, when properly performed, the most accurate method available for determining surface tensions. Since the measurements do not involve a disturbance of the surface, slow time effects can be followed.

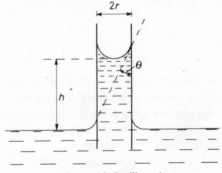

Figure 4.2. Capillary rise

For the rise of a liquid up a narrow capillary

$$\gamma = \frac{rh\Delta\rho g}{2\cos\theta} \qquad \dots (4.3)$$

which for zero contact angle reduces to

$$\gamma = \tfrac{1}{2}rh\Delta\rho g \qquad \dots (4.4)$$

where $\Delta\rho$ is (density of liquid $-$ density of vapour).

For accurate work a meniscus correction should be made. In a narrow capillary the meniscus will be approximately hemispherical; therefore

$$\gamma = \tfrac{1}{2}r\,(h+r/3)\,\Delta\rho g \qquad \dots (4.5)$$

For wider capillaries one must account for deviation of the meniscus from hemispherical shape.[3]

In practice, the capillary rise method is only used when the contact angle is zero, owing to the uncertainty in measuring contact angles correctly. One can check for zero contact angle by allowing

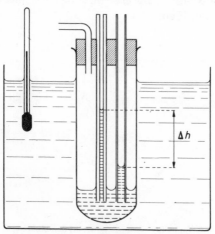

Figure 4.3. Differential capillary rise apparatus

the meniscus to approach an equilibrium position in turn from above and below. With a finite contact angle, different equilibrium positions will be noted due to differences between advancing and receding contact angles (see page 118). Zero contact angle for aqueous and most other liquids can usually be obtained without difficulty using well-cleaned glass capillaries. (The need for extreme cleanliness in surface chemical experiments cannot be overemphasised.)

A difficulty associated with this method is that of obtaining capillary tubing of uniform bore (slight ellipticity is not important). Thermometer tubing is useful in this respect. Alternatively, one can adjust the height of the reservoir liquid so as to locate the meniscus at a particular level in the capillary where the cross-sectional area is accurately known.

A useful variation is to measure the difference in capillary rise for capillaries of different bore, thus eliminating reference to the flat surface of the reservoir liquid (Figure 4.3). Since

$$\gamma = \tfrac{1}{2}r_1 h_1 \Delta \rho g = \tfrac{1}{2}r_2 h_2 \Delta \rho g$$

then
$$\gamma = \frac{\Delta \rho g r_1 r_2 \Delta h}{2(r_1 - r_2)} \qquad \ldots \ldots (4.6)$$

2. *Wilhelmy plate methods*—A thin mica plate or microscope slide is suspended from the arm of a balance and dips into the liquid, as shown in Figure 4.4.

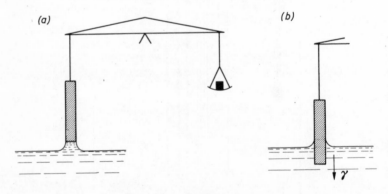

Figure 4.4. Wilhelmy plate methods: (*a*) detachment; (*b*) static

(a) When used as a detachment method (Figure 4.4a), the container holding the liquid is gradually lowered and the pull on the balance at the point of detachment is noted. For a plate of length x, breadth y and weight W, assuming zero contact angle

$$W_{\text{det.}} - W = 2(x+y)\gamma \qquad \ldots \ldots (4.7)$$

(b) The plate method can also be used as a static method (Figure 4.4b) for measuring changes in surface tension (see film balance, page 82). The change in the force required to maintain the plate at constant immersion as the surface tension alters is measured.

3. *Ring method*— In this method the force required to detach a ring from a surface or interface is measured either by suspending

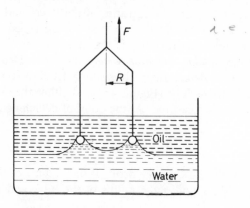

Figure 4.5. Measurement of interfacial tension by the ring method

the ring from the arm of a balance or by using a torsion wire arrangement (du Noüy tensiometer). The detachment force is related to the surface or interfacial tension by the expression

$$\gamma = \frac{\beta F}{4\pi R} \qquad \qquad \ldots . (4.8)$$

where F is the pull on the ring, R the mean radius of the ring, and β a correction factor.

To ensure zero, and hence a constant contact angle, platinum rings carefully cleaned in strong acid or by flaming are used. It is essential that the ring should lie flat in a quiescent surface. For interfacial work, the lower liquid must wet the ring preferentially (e.g. for benzene on water, a clean platinum ring is suitable; whereas for water on carbon tetrachloride, the ring must be made hydrophobic).

The correction factor β allows for the non-vertical direction of the tension forces and for the complex shape of the liquid supported by the ring at the point of detachment; hence it depends on the dimensions of the ring and the nature of the interface. Values of β have been tabulated by Harkins and Jordan,[46] they can also be calculated from the equation of Zuidema and Waters[47]

$$(\beta - a)^2 = \frac{4b}{\pi^2} \cdot \frac{1}{R^2} \cdot \frac{F}{4\pi R (\rho_1 - \rho_2)} + c \qquad \dots (4.9)$$

where ρ_1 and ρ_2 are the densities of the lower and upper phases; $a = 0.7250$ and $b = 0.09075 \text{ m}^{-1} \text{ s}^2$ for all rings; $c = 0.04534 - 1.679 \, r/R$; and r is the radius of the wire.

4.   *Drop volume and drop weight methods*—Drops of a liquid are allowed to detach themselves slowly from the tip of a vertically mounted narrow tube (Figure 4.6) and either they are weighed or their volume is measured. At the point of detachment

$$\gamma = \frac{\phi mg}{2\pi r} = \frac{\phi V \rho g}{2\pi r} \qquad \dots (4.10)$$

where m is the mass of the drop, V the volume of the drop, ρ the density of the liquid, r the radius of the tube, and ϕ a correction factor.

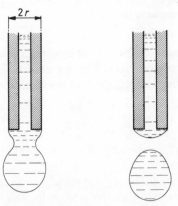

Figure 4.6. Detachment of a drop from the tip of a narrow tube

The correction factor ϕ is required because on detachment (*a*) the drop does not completely leave the tip and (*b*) the surface tension forces are seldom exactly vertical. ϕ depends on the ratio $r/V^{1/3}$. Values of ϕ have been determined empirically by Harkins and

Brown.[48, 49] It can be seen that values of $r/V^{1/3}$ between about 0·6 and 1·2 are preferable (Figure 4.7).

A suitable tip which has been carefully ground smooth used in conjunction with a micrometer syringe burette gives a very convenient drop volume apparatus for determining both surface and

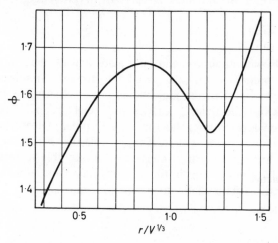

Figure 4.7. Correction factor for drop volume and drop weight methods

interfacial tensions. The tip of the tube must be completely wetted (r = external radius); alternatively, a tip with a sharp edge can be used. For accurate measurements, the set-up should be free from vibration and the last 10 per cent of the drop should be formed very slowly ($\sim$1 minute).

5.   *Pendant drop method*—A pendant drop of liquid is photographed, or its image projected on to graph paper. From the various dimensions of the drop the surface or interfacial tension can be computed.[3]

6.   *Oscillating jet method*—This is a dynamic method which enables one to measure the tensions of surfaces at very short times (*ca.* 0·01 s) from the moment of their creation. (The methods previously described are used to measure equilibrium tensions.) A jet of liquid emerging from a nozzle of elliptical cross-section is unstable and oscillates about its preferred circular cross-section. Surface

tensions can be calculated from the jet dimensions (obtained photo-graphically), flow rate, etc. The age of the surface can be controlled to some extent by altering the flow rate.

ADSORPTION AND ORIENTATION AT INTERFACES

Surface activity

Materials such as short-chain fatty acids and alcohols are soluble in both water and oil (e.g. paraffin hydrocarbon) solvents. The hydrocarbon part of the molecule is responsible for its solubility in oil, whilst the polar —COOH or —OH group has sufficient affinity to water to drag a short-length non-polar hydrocarbon chain into aqueous solution with it. If these molecules become located at an air–water or oil–water interface, they are able to locate their hydrophilic head groups in the aqueous phase and allow the lipo-philic hydrocarbon chains to escape into the vapour or oil phase

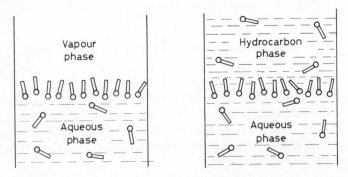

Figure 4.8. Adsorption of surface-active molecules as an orientated mono-layer at air–water and oil–water interfaces. The circular part of the molecules represents the hydrophilic polar head group and the rectangular part represents the non-polar hydrocarbon tail

(Figure 4.8). This situation is energetically more favourable than complete solution in either phase.

The strong adsorption of such materials at surfaces or interfaces in the form of an orientated *monomolecular layer* (or *monolayer*) is termed *surface activity*. Surface-active materials (or *surfactants*) consist of molecules containing both polar and non-polar parts (*amphiphilic*). Surface activity is a dynamic phenomenon, since the final state of a surface or interface represents a balance between this tendency towards adsorption and the tendency towards complete mixing due to the thermal motion of the molecules.

The tendency for surface-active molecules to pack into an interface favours an expansion of the interface; this must therefore be

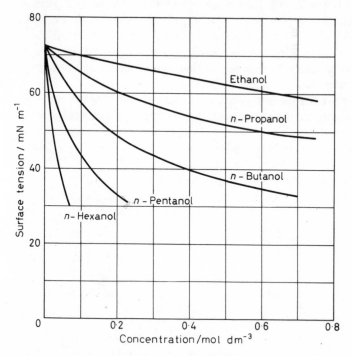

Figure 4.9. Surface tension of aqueous solutions of alcohols at 20°C

balanced against the tendency for the interface to contract under normal surface tension forces. If π is the expanding pressure (or *surface pressure*) of an adsorbed layer of surfactant, then the surface (or interfacial) tension will be lowered to a value

$$\gamma = \gamma_0 - \pi \qquad \ldots (4.11)$$

Figure 4.9 shows the effect of lower members of the homologous series of normal fatty alcohols on the surface tension of water. The longer the hydrocarbon chain, the greater is the tendency for the alcohol molecules to adsorb at the air–water surface and hence lower the surface tension. A rough generalisation, known as Traube's rule, is that for a particular homologous series of surfactants the concentration required for an equal lowering of surface tension in dilute solution decreases by a factor of about three for each additional CH_2 group.

If the interfacial tension between two liquids is reduced to a sufficiently low value on addition of a surfactant, emulsification will readily take place because only a relatively small increase in the surface free energy of the system is involved. If $\pi \geqslant \gamma_0$, either the liquids will become miscible or spontaneous emulsification will occur.

In certain cases—solutions of electrolytes, sugars, etc.—small increases in surface tension due to negative adsorption are noted.

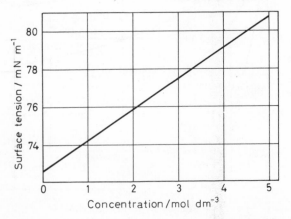

Figure 4.10. Surface tension of aqueous sodium chloride solutions at 20°C

Here, because the solute–solvent attractive forces are greater than the solvent–solvent attractive forces, the solute molecules tend to migrate away from the surface into the bulk of the liquid.

Classification of surfactants

The hydrophilic part of the most effective soluble surfactants (e.g. soaps, synthetic detergents and dyestuffs) is often an ionic group.

Ions have a strong affinity for water owing to their electrostatic attraction to the water dipoles and are capable of pulling fairly long hydrocarbon chains into solution with them; for example, palmitic acid, which is virtually unionised, is insoluble in water, whereas sodium palmitate, which is almost completely ionised, is soluble (especially above is Krafft temperature—see page 77).

Surfactants are classified as *anionic, cationic* or *non-ionic*[52] according to the charge carried by the surface-active part of the molecule. Some common examples are given in Table 4.2.

Table 4.2   SURFACE-ACTIVE AGENTS

Anionic	
Sodium stearate	$CH_3(CH_2)_{16}COO^-Na^+$
Sodium oleate	$CH_3(CH_2)_7CH = CH(CH_2)_7COO^-Na^+$
Sodium dodecyl sulphate	$CH_3(CH_2)_{11}SO_4^-Na^+$
Sodium dodecyl benzene sulphonate	$CH_3(CH_2)_{11}.C_6H_4.SO_3^-Na^+$
Cationic	
Laurylamine hydrochloride	$CH_3(CH_2)_{11}NH_3^+Cl^-$
Cetyl trimethyl ammonium bromide	$CH_3(CH_2)_{15}N(CH_3)_3^+Br^-$
Non-ionic	
Polyethylene oxides e.g.	$CH_3(CH_2)_7.C_6H_4.(O.CH_2.CH_2)_8OH$
Spans (sorbitan esters)	
Tweens (polyoxyethylene sorbitan esters)	

Rate of adsorption

The formation of an adsorbed surface layer is not an instantaneous process but is governed by the rate of diffusion of the surfactant through the solution to the interface. It might take several seconds for a surfactant solution to attain its equilibrium surface tension, especially if the solution is dilute and the solute molecules are large and unsymmetrical. Much slower ageing effects have been reported, but these are now known to be due to traces of impurities. The time factor in adsorption can be demonstrated by measuring the surface tensions of freshly formed surfaces by a dynamic method; for example, the surface tensions of sodium oleate solutions measured by the oscillating jet method approach that of pure water but fall rapidly as the surfaces are allowed to age.

Thermodynamics of adsorption—Gibbs adsorption equation

The Gibbs adsorption equation enables the extent of adsorption at a liquid surface to be estimated from surface tension data.

The quantitative treatment of surface phenomena involves an important uncertainty. It is convenient to regard the interface between two phases as a mathematical plane, such as SS in Figure 4.11. This approach, however, is unrealistic, especially if an adsorbed film is present. Not only will such a film itself have a certain thickness, but its presence may influence nearby structure (for example, by dipole–dipole orientation—especially in an aqueous phase) and

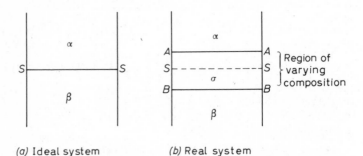

(a) Ideal system   *(b)* Real system

Figure 4.11. Representations of an interface between bulk phases α and β

result in an interfacial region of varying composition with an appreciable thickness in terms of molecular dimensions.

If a mathematical plane is, nevertheless, taken to represent the interface between two phases, adsorption can be described conveniently in terms of *surface excess concentrations*. If n_i^σ is the amount of component i in the surface phase σ (Figure 4.11b) in excess of that which would have been in σ had the bulk phases α and β extended to a surface SS with unchanging composition, the surface excess concentration of component i is given by

$$\Gamma_i = \frac{n_i^\sigma}{A} \qquad \qquad \ldots \ldots (4.12)$$

where A is the interfacial area. Γ_i may be positive or negative, and its magnitude clearly depends on the location of SS which (as illustrated in the following derivation) must be chosen somewhat arbitrarily.

The total thermodynamic energy of a system is given by the expression

$$U = TS - pV + \Sigma \mu_i n_i$$

The corresponding expression for the thermodynamic energy of a surface phase σ is

$$U^\sigma = TS^\sigma - pV^\sigma + \gamma A + \Sigma \mu_i n_i^\sigma \qquad \dots (4.13)$$

(The pV^σ and γA terms have opposite signs, since pressure is an expanding force and surface tension is a contracting force. A superscript is not necessary for T, p and the chemical potential terms, since these must have uniform values throughout for a heterogeneous system to be in equilibrium.) Differentiating Equation (4.13) generally

$$dU^\sigma = TdS^\sigma + S^\sigma dT - pdV^\sigma - V^\sigma dp + \gamma dA + Ad\gamma$$
$$+ \Sigma \mu_i \, dn_i^\sigma + \Sigma n_i^\sigma \, d\mu_i \qquad \dots (4.14)$$

From the first and second laws of thermodynamics

$$dU = TdS - pdV + \Sigma \mu_i \, dn_i$$

or, for a surface phase

$$dU^\sigma = TdS^\sigma - pdV^\sigma + \gamma dA + \Sigma \mu_i \, dn_i^\sigma \qquad \dots (4.15)$$

Subtracting Equation (4.15) from Equation (4.14)

$$S^\sigma dT - V^\sigma dp + Ad\gamma + \Sigma n_i^\sigma \, d\mu_i = 0$$

Therefore, at constant temperature and pressure

$$d\gamma = -\sum \frac{n_i^\sigma}{A} \, d\mu_i = -\sum \Gamma_i \, d\mu_i \qquad \dots (4.16)$$

For a simple two-component solution (i.e. consisting of a solvent and a single solute) Equation (4.16) becomes

$$d\gamma = -\Gamma_A \, d\mu_A - \Gamma_B \, d\mu_B$$

As explained above, surface excess concentrations are defined relative to an arbitrarily chosen dividing surface. A convenient (and seemingly realistic) choice of location of this surface for a binary solution is that at which the surface excess concentration of the solvent (Γ_A) is zero. The above expression then simplifies to

$$d\gamma = -\Gamma_B \, d\mu_B$$

Since chemical potential changes are related to relative activities by

$$\mu_B = \mu_B^\ominus + RT \ln a_B$$

then

$$d\mu_B = RT \, d \ln a_B$$

Therefore

$$\Gamma_B = -\frac{1}{RT} \cdot \frac{d\gamma}{d \ln a_B} = -\frac{a_B}{RT} \cdot \frac{d\gamma}{da_B} \qquad \ldots (4.17)$$

or, for dilute solutions,

$$\Gamma_B = -\frac{c_B}{RT} \cdot \frac{d\gamma}{dc_B} \qquad \ldots (4.18)$$

which is the form in which the Gibbs equation is usually quoted.

Experimental verification of the Gibbs equation

The general form of the Gibbs equation ($d\gamma = -\Sigma \, \Gamma_i \, d\mu_i$) is fundamental to all adsorption processes. However, experimental verifica-
tion of the equation derived for simple systems is of interest in view
of the postulation which was made concerning the location of the
boundary surface. Several early investigators bubbled air through
solutions of surface-active materials and collected the froth which
was formed. The surface excess was calculated via an analysis of
the collapsed froth and an estimation of the surface area of the
bubbles in the froth. However, the results were unreliable owing
to the uncertainty in estimating this area and because adsorption
equilibrium was not fully established.

McBain et al.[50] succeeded in verifying the validity of the Gibbs
equation by means of a very direct and ingenious experiment.
Surface layers of about 0·1 mm thickness were shaved off solutions
of surface-active materials, such as phenol and hydrocinnamic acid,
contained in a long rectangular trough, by means of a rapidly
moving microtome blade. The material collected was analysed and
experimental surface excess concentrations calculated. These compared well with the corresponding surface excess concentrations
calculated from surface tension data.

Surface concentrations have also been successfully measured[51]
by labelling the solute with a β-emitting radioactive isotope and
measuring the radiation picked up by a Geiger counter placed
immediately above the surface of the solution. As β-rays are rapidly

attenuated in the solution, the measured radiation corresponds to the surface region plus only a thin layer of bulk region. Direct measurement of surface concentrations is particularly useful when there is more than one surface-active species or unavoidable surface-active impurity present. In such cases surface tension measurements would probably be ambiguous.

ASSOCIATION COLLOIDS

Physical properties of surfactant solutions—micelle formation

Solutions of highly surface-active materials exhibit unusual physical properties. In dilute solution they act as normal electrolytes, but

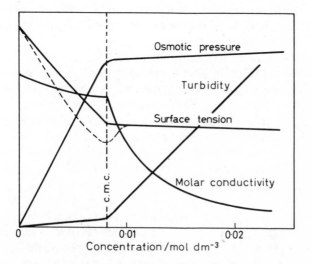

Figure 4.12. Physical properties of sodium dodecyl sulphate solutions at 25°C

at fairly well defined concentrations abrupt changes in several physical properties such as osmotic pressure, conductance, turbidity and surface tension take place. The rate at which osmotic pressure increases with concentration becomes abnormally low, suggesting that considerable association is taking place; yet the conductance of ionic surfactant solutions remains relatively high, showing that ionic dissociation is still in force.

McBain pointed out that this seemingly anomalous behaviour could be explained in terms of organised aggregates, or *micelles*, of the surfactant ions in which the lipophilic hydrocarbon chains are orientated towards the interior of the micelle, leaving the hydrophilic groups in contact with the aqueous medium. The concentration above which micelle formation becomes appreciable is termed the *critical micelle concentration* (c.m.c.).

Micellisation is, therefore, an alternative mechanism to adsorption by which the interfacial energy of a surfactant solution might decrease. Thermal agitation and electrostatic repulsion between the charged head groups on the surface of the micelle oppose this aggregating tendency. Consequently, a low c.m.c. would be expected to be favoured by:

1. Increasing the hydrophobic part of the surfactant molecules. (For a homologous series each additional CH_2 group approximately halves the c.m.c.)
2. Lowering the temperature.
3. The addition of simple salts (e.g. KCl) which reduce the above repulsive forces by their screening action (see Chapter 7).

Many non-ionic surfactants also form micelles, often at very small concentrations (*ca.* 10^{-4} mol dm^{-3}).

Structure of micelles

Micellar theory has developed in a somewhat uncertain fashion and is still in some respects open to discussion. Hartley[53] proposed a spherical shape and suggested that micelles are essentially liquid droplets of colloidal dimensions with the charged groups situated at the surface; most of the available evidence favours this model. In addition, McBain[12] believed that a laminar form also exists, and Harkins[10] considered the possibility of a cylindrical micelle.

Some of the experimental evidence favouring the existence of spherical liquid micelles is summarised as follows:

1. Critical micelle concentrations depend almost entirely on the nature of the lyophobic part of the surfactant. If micelle structure involved some kind of crystal lattice arrangement

then the nature of the lyophilic head group would also be expected to be important.

2. Micelles have a definite fixed size, which again depends almost entirely on the nature of the lyophobic part of the surfactant molecules. One would expect the radius of spherical micelles to be slightly less than the length of the constituent units, otherwise the hydrocarbon chains would be considerably buckled or the micelle would have either a hole or ionic groups in the centre. The radii of micelles calculated from diffusion and light-scattering data support this expectation.

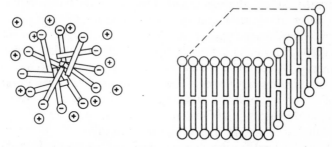

Figure 4.13. Schematic representation of spherical and laminar micelle structures

For straight-chain surfactants the number of monomer units per micelle, *m*, and the number of carbon atoms per hydrocarbon chain, *n*, are approximately related as follows:[54]

$n =$	12	$m =$	33
	14		46
	16		60
	18		78

$\frac{n}{m} = 0.3 \to 0.2$

3. Surfactant solutions above the c.m.c. can solubilise otherwise insoluble organic material by incorporating it into the interior of the micelles;[55] for example, the dye xylenol orange dissolves only sparingly in pure water but gives a deep red solution with sodium dodecyl sulphate. For diffusion reasons, solubilisation would not take place readily if the micelle were solid.

The laminar micelle theory supposes that the surfactant molecules are organised in the form of a double layer (Figure 4.13), with the head groups facing outwards. This model gained favour for a time, mainly on the basis of x-ray evidence. When a beam of mono-chromatic x-rays was passed through a thin layer of soap solution, diffraction patterns were observed which, on interpretation via the Bragg equation, gave spacings which were consistent with those expected in a laminar micelle. On adding oil-soluble material such as benzene, an increase in the long spacing was observed which was consistent with solubilisation in the form of a thin film between the layers of the micelle.

However, the laminar theory is inconclusive in that it provides no ready mechanism by which the size of the micelles might be limited; in fact, thermodynamically one would expect that laminar micelles should tend to grow to an infinite size in order to minimise the interfacial energy at the hydrocarbon edges. The spherical micelle permits an almost complete elimination of this interfacial energy. Moreover, it is now realised that the above x-ray evidence is ambiguous since it can also be attributed to a certain amount of long-range ordering of spherical micelles caused by electrostatic repulsive forces. If laminar (or cylindrical) micelles do exist, they probably do so only in concentrated solutions ($>$ ca. 10 per cent); streaming birefringence evidence suggests that this might be the case.

Surface behaviour

Figure 4.12 shows how a highly surface-active material such as sodium dodecyl sulphate lowers the surface tension of water quite appreciably even at low concentrations. The discontinuity in the γ-composition curve is identified with the c.m.c., beyond which there is an additional mechanism for keeping hydrocarbon chains away from water surfaces, i.e. by locating them in the interior of the micelles. Since the micelles themselves are not surface-active the surface tension remains approximately constant beyond the c.m.c. The minimum in the γ-composition curve, shown by the dashed curve, is typical of measurements which have been made on colloidal electrolyte solutions and in apparent violation of the Gibbs equation, since it suggests desorption over the small concentration range where $d\gamma/dc$ is positive. This anomaly is attributed to traces of impurity such as dodecanol, which is surface-adsorbed below the c.m.c. but solubilised by the micelles beyond the c.m.c. With sufficient purification the minimum in the γ-composition curve can be removed. Beyond the c.m.c., where $d\gamma/dc \sim 0$, application of the Gibbs equation might suggest almost zero adsorption; however, $d\gamma/da$, where a represents the activity of single surfactant species, is still appreciably negative, a changing little above the c.m.c.

Conductance

Micelle formation affects the conductance of ionic surfactant solutions for the following reasons:

1. The total viscous drag on the surfactant molecules is reduced on aggregation.
2. Counter-ions become kinetically a part of the micelle owing to its high surface charge (see Chapter 7), thus reducing the number of counter-ions available for carrying the current and also lowering the net charge of the micelles.
3. The retarding influence of the ionic atmospheres of un-attached counter-ions on the migration of the surfactant ions is greatly increased on aggregation.

The last two factors, which cause the molar conductivity to decrease with concentration beyond the c.m.c., normally outweigh the first factor, which has the reverse effect (see Figure 4.12). When conductance measurements are made at very high field strengths the ionic atmospheres cannot re-form quickly enough (Wien effect) and some of the bound counter-ions are set free. It is interesting to note that under these conditions the molar conductivity increases with concentration beyond the c.m.c.

Energetics of micellisation

In general, micellisation is an exothermic process and is, therefore, favoured by a decrease in temperature (see page 72). This, however, is not universally the case; for example, the c.m.c. of sodium dodecyl sulphate in water shows a shallow minimum between about 20°C and 25°C. At lower temperature the enthalpy of micellisation given from the expression

$$\frac{d \ln (\text{c.m.c.})}{dT} = \frac{\Delta H_{\text{micellisation}}}{2RT^2}$$

is positive (endothermic), and micellisation is entirely entropy-directed.

The cause of a positive entropy of micellisation is not entirely clear. A decrease in the amount of water structure as a result of micellisation may make some contribution. A more likely contribution, however, involves the configuration of the hydrocarbon chains which probably have considerably more freedom of movement in the interior of the micelle than when in contact with the aqueous medium.

Sharpness of critical micelle concentrations

There are two current theories relating to the abruptness with which micellisation takes place above a certain critical concentration.[56, 57]

The first of these theories applies the law of mass action to the equilibrium between unassociated molecules or ions and micelles, as illustrated by the following simplified calculation. If c is the stoichiometric concentration of the solution, x the fraction of monomer units aggregated and m the number of monomer units per micelle

$$\underset{c\,(1-x)}{mX} = \underset{cx/m}{(X)_m}$$

Therefore, applying the law of mass action

$$K = \frac{cx/m}{[c\,(1-x)]^m} \qquad \qquad \text{.... (4.19)}$$

For moderately large values of m, this expression requires that x should remain very small up to a certain value of c and increase rapidly thereafter. The sharpness of the discontinuity will depend on the value of m ($m = \infty$ gives a perfect discontinuity). If this treatment is modified to account for the counter-ions associated with an ionic micelle, then an even more abrupt discontinuity than the above is predicted.

The alternative approach is to treat micellisation as a simple phase separation of surfactant in an associated form, with the unassociated surfactant concentration remaining practically constant above the c.m.c.

The Krafft phenomenon

Micelle-forming surfactants exhibit another unusual phenomenon in that their solubilities show a rapid increase above a certain temperature, known as the *Krafft point*. The explanation of this behaviour arises from the fact that unassociated surfactant has a limited solubility, whereas the micelles are highly soluble. Below the Krafft temperature the solubility of the surfactant is insufficient for micellisation. As the temperature is raised the solubility slowly increases until, at the Krafft temperature, the c.m.c. is reached. A relatively large amount of surfactant can now be dispersed in the form of micelles, so that a large increase in solubility is observed.

SPREADING

Adhesion and cohesion

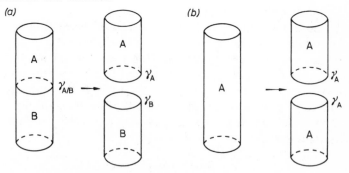

Figure 4.14. Work of adhesion (*a*) and work of cohesion (*b*)

(a) The *work of adhesion* between two immiscible liquids is equal to the work required to separate unit area of the liquid–liquid interface and form two separate liquid–air interfaces (Figure 4.14a), and is given by the Dupré equation

$$W_a = \gamma_A + \gamma_B - \gamma_{A/B} \qquad \qquad \ldots\ldots (4.20)$$

(b) The *work of cohesion* for a single liquid corresponds to the work required to pull apart a column of liquid of unit cross-sectional area (Figure 4.14b), i.e.

$$W_c = 2\gamma_A \qquad \qquad \ldots\ldots (4.21)$$

Spreading of one liquid on another

When a drop of an insoluble oil is placed on a clean water surface it may behave in one of three ways:

1. Remain as a lens, as in Figure 4.15 (non-spreading).

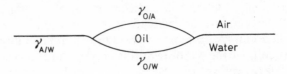

Figure 4.15. A drop of non-spreading oil on a water surface

2. Spread as a thin film, which may show interference colours, until it is uniformly distributed over the surface as a 'duplex' film. (A duplex film is a film which is thick enough for the two interfaces, i.e. liquid–film and film–air, to be independent and possess characteristic surface tensions.)
3. Spread as a monolayer, leaving excess oil as lenses in equilibrium, as in Figure 4.16.

If the area occupied by the oil drop shown in Figure 4.15 is increased by dA, the change in the surface free energy of the system will be approximately $(\gamma_{O/A} + \gamma_{O/W} - \gamma_{W/A})\,dA$. If this quantity is negative, the process of spreading will take place spontaneously.

Harkins defined the term *initial spreading coefficient* (for the case of oil on water) as

$$S = \gamma_{W/A} - (\gamma_{O/A} + \gamma_{O/W}) \qquad \dots (4.22)$$

where the various interfacial tensions are measured before mutual saturation of the liquids in question has occurred. The condition for initial spreading is therefore that S be positive or zero (Table 4.3).

Substituting in the Dupré equation, the spreading coefficient can be related to the work of adhesion and cohesion

$$S = W_{O/W} - 2\gamma_{O/A} = W_{O/W} - W_{Oil} \qquad \dots (4.23)$$

i.e. spreading occurs when the oil adheres to the water more strongly than it coheres to itself.

Impurities in the oil phase, e.g. oleic acid in hexadecane, can reduce $\gamma_{O/W}$ sufficiently to make S positive. Impurities in the aqueous phase normally reduce S, since $\gamma_{W/A}$ is lowered more than $\gamma_{O/W}$ by the impurity, especially if $\gamma_{O/W}$ is already low. Therefore, n-octane will spread on a clean water surface but not on a contaminated surface.

The initial spreading coefficient does not consider the mutual saturation of one liquid with another: for example, when benzene is spread on water

$$S_{\text{init.}}/\text{mN m}^{-1} = 72\cdot8 - (28\cdot9 + 35\cdot0) = +8\cdot9$$

but when the benzene and water have had time to become mutually saturated, $\gamma_{W/A}$ is reduced to $62\cdot4 \text{ mN m}^{-1}$ and $\gamma_{O/W}$ to $28\cdot8 \text{ mN m}^{-1}$ so that

$$S_{\text{final}}/\text{mN m}^{-1} = 62\cdot4 - (28\cdot8 + 35\cdot0) = -1\cdot4$$

Table 4.3 INITIAL SPREADING COEFFICIENTS (IN mN m^{-1}) FOR LIQUIDS ON WATER AT 20°C[5]
(By courtesy of Academic Press Inc.)

Liquid	$\gamma_{W/A} - (\gamma_{O/A} + \gamma_{O/W}) = S$	Conclusion
n-Hexadecane	$72\cdot8 - (30\cdot0 + 52\cdot1) = \ -9\cdot3$	will not spread on water
n-Octane	$72\cdot8 - (21\cdot8 + 50\cdot8) = \ +0\cdot2$	will just spread on pure water
n-Octanol	$72\cdot8 - (27\cdot5 + \ 8\cdot5) = +36\cdot8$	will spread against contamination

The final state of the interface is now just unfavourable for spreading. This causes the initial spreading to be stopped, and can even

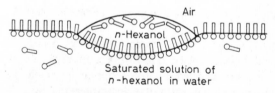

Figure 4.16. Spreading of n-hexanol on a water surface

result in the film retracting slightly to form very flat lenses, the rest of the water surface being covered by a monolayer of benzene.

Similar considerations apply to the spreading of a liquid such as n-hexanol on water (Figure 4.16).

$$S_{\text{init.}}/\text{mN m}^{-1} = 72\cdot8 - (24\cdot8 + 6\cdot8) = +41\cdot2$$
$$S_{\text{final}}/\text{mN m}^{-1} = 28\cdot5 - (24\cdot7 + 6\cdot8) = -\ 3\cdot0$$

MONOMOLECULAR FILMS[2, 58]

Many insoluble substances, such as long-chain fatty acids and alcohols, can (with the aid of suitable solvents) be spread on to a water surface, and if space permits will form a surface film one molecule in thickness with the hydrophilic —COOH or —OH groups orientated towards the water phase and the hydrophobic hydrocarbon chains orientated away from the water phase.

These insoluble monomolecular films, or monolayers, represent an extreme case in adsorption at liquid surfaces as all the molecules in question are concentrated in one molecular layer at the interface. In this respect they lend themselves to direct study. In contrast to monolayers which are formed by adsorption from solution, the surface concentrations of insoluble films are known directly from the amount of material spread and the area of the surface, recourse to the Gibbs equation being unnecessary.

The molecules in a monomolecular film, especially at high surface concentrations, are often arranged in a simple manner and much can be learned about the size, shape and orientation of the individual molecules by studying various properties of the monolayer. Monomolecular films can exist in different, two-dimensional physical states, depending mainly on the magnitude of the lateral adhesive forces between the film molecules, in much the same way as three-dimensional matter.

Experimental techniques for studying insoluble monolayers

1. *Surface pressure*—The surface pressure of a monolayer is the lowering of surface tension due to the monolayer, i.e. it is the expanding pressure exerted by the monolayer which opposes the normal contracting tension of the clean interface, or

$$\pi = \gamma_0 - \gamma \qquad \qquad \ldots (4.11)$$

where γ_0 is the tension of clean interface, and γ the tension of interface plus monolayer.

The variation of surface pressure with the area available to the spread material is represented by a $\pi - A$ (force–area) curve. With a little imagination, $\pi - A$ curves can be regarded as the two-dimensional equivalent of the $p - V$ curves for three-dimensional matter. (N.B. For a 1 nm thick film, a surface pressure of $1 \, \text{mN m}^{-1}$ is equivalent to a bulk pressure of $10^6 \, \text{N m}^{-2}$, or ~ 10 atmospheres.)

The Langmuir-Adam surface balance (or trough) uses a technique for containing and manoeuvring insoluble monolayers between barriers for the direct determination of $\pi - A$ curves. The film is contained (Figure 4.17) between a movable barrier and a float attached to a torsion wire arrangement. The surface pressure of the film is measured directly in terms of the horizontal force which it exerts on the float and the area of the film is varied by means of the movable barrier.

The film must be contained entirely between the barrier and the float without any leakage. To achieve this, the trough walls, barrier and float must be hydrophobic and the liquid level must be slightly

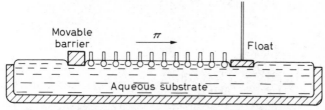

Figure 4.17. The principle of the Langmuir-Adam surface balance

above the brim of the trough. Teflon troughs and accessories are ideal in this respect. Silica or glass apparatus which has been made hydrophobic with a light coating of purified paraffin wax or silicone waterproofing material is also satisfactory. Waxed threads prevent leakage past the ends of the float.

Before spreading, the surface must be carefully cleaned and freed from contamination. Redistilled water should be used. The liquid surface in front of and behind the float can be swept clean by moving barriers towards the float and sucking away any surface impurities with a capillary joined to a water pump.

To achieve uniform spreading the material in question is normally predissolved in a solvent such as petroleum ether to give a *ca.* 0·1 per cent solution. A total of *ca.* 0·01 cm^3 of this spreading solution is ejected in small amounts from a micrometer syringe burette at various points on the liquid surface. The spreading solvent evaporates away leaving a uniformly spread film. Benzene, although frequently used as a spreading solvent, is not entirely suitable owing to its slight solubility in water and long residence time at the interface (see page 79).

The surface pressure of the film is determined by measuring the force which must be applied via a torsion wire to maintain the

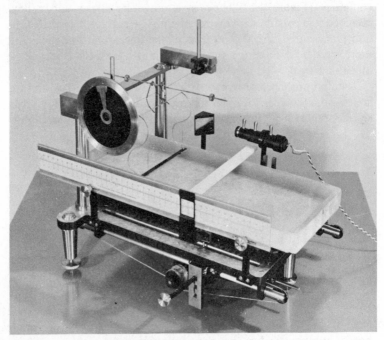

Figure 4.18. A surface balance (By courtesy of Unilever Research Laboratory, Port Sunlight)

float at a fixed position on the surface (located optically) and dividing by the length of the float. For precise work, the surface balance is enclosed in an air thermostat and operated by remote control. With a good modern instrument, surface pressures can be measured with an accuracy of 0.01 mN m^{-1}

Surface pressures can also be determined indirectly by measuring the lowering of the surface tension of the underlying liquid (or substrate) caused by the film. For example, the float can be replaced by a Wilhelmy plate set-up (Figure 4.4b). This method is at least as accurate as the Langmuir-Adam surface balance and is particularly useful for studying films at oil–water interfaces. The obvious advantage of this method is that it is not necessary to maintain a clean surface behind a float. The arrangement can be simplified even further by doing away with the movable barrier as well, and regulating the area of the film by the amount of material spread.

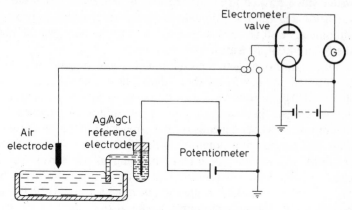

Figure 4.19. Air electrode method

2. *Surface film potential*—In heterogeneous systems, potential differences exist across the various phase boundaries. The *surface film potential, ΔV,* due to a monolayer is the change in the potential

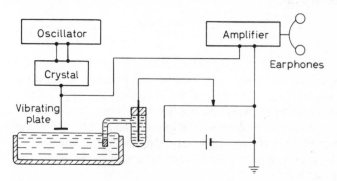

Figure 4.20. Vibrating plate method

difference between a liquid and a probe placed above the liquid surface arising from the presence of the monolayer. Surface film potentials are measured by either the air electrode (Figure 4.19) or vibrating plate method (Figure 4.20).

The air electrode consists of an insulated metal wire which is mounted with its tip 1 mm to 2 mm above the water surface. Polonium deposited on the tip of the electrode renders the air gap conducting. The galvanometer reading with the grid of the electrometer valve earthed is noted. The grid is then connected to the air electrode and the potentiometer adjusted to give the original galvanometer reading. This is repeated with the spread film, ΔV being the difference in potentiometer readings.

In the vibrating plate method, a small gold or gold-plated disc is mounted about 0·5 mm above the surface. The vibration of the disc ($\sim$200 Hz) produces a corresponding variation in the capacity across the air gap, thus setting up an alternating current, the magnitude of which depends on the potential difference across the gap. The surface film potential is the difference in potentiometer readings required to give minimum sound in the ear-phones for clean and film-covered surfaces. This method is more accurate than the air electrode method, being capable of measuring ΔV to 0·1 mV, but is subject to malfunctioning. It can also be used at oil–water interfaces.

Surface film potential measurements can yield useful, if not absolute, information about the orientation of the film molecules. Treating the film as a parallel plate condenser leads to the approximate expression

$$\Delta V = \frac{n\mu \cos \theta}{\varepsilon} \qquad \qquad \ldots \ . \ (4.24)$$

where n is the number of film molecules per unit area, μ the dipole moment of film molecules, θ the angle of inclination of dipoles to the normal, and ε the permittivity of the film (see page 137).

Surface film potential measurements are also used to investigate the homogeneity or otherwise of the surface. If there are two surface phases present, the surface film potential will fluctuate wildly as the probe is moved across the surface or as one blows gently over the surface.

3. *Surface rheology*—Surface viscosity is the change in the viscosity of the surface layer brought about by the monomolecular film. Monolayers in different physical states can be readily distinguished by surface viscosity measurements.

A qualitative idea of the surface viscosity is given by noting the ease with which talc can be blown about the surface. Most insoluble films have surface viscosities of *ca.* 10^{-6} kg s^{-1} to 10^{-3} kg s^{-1} (for films 10^{-9} m thick this is equivalent to a bulk viscosity range of *ca.* 10^3 kg m^{-1} s^{-1} to 10^6 kg m^{-1} s^{-1}). These films can be studied by means of a damped oscillation method (Figure 4.21). For a vane of length *l* and a disc with a moment of inertia *I*

$$\eta_s = \frac{9\cdot 2\,I}{l^2}\left(\frac{\lambda}{t} - \frac{\lambda_0}{t_0}\right) \qquad \dots\,(4.25)$$

where $\lambda = \log_{10}$ of the ratio of successive amplitudes of damped

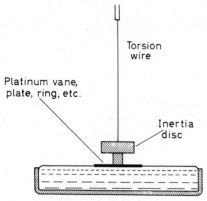

Figure 4.21. Damped oscillation method for measuring surface viscosities

oscillation, *t* is the period of oscillation, and the subscripts '0' refer to a clean surface.

Many insoluble films, particularly those containing protein, exhibit viscoelastic behaviour (see Chapter 9). A surface rheometer has been designed[59] for studying film creep under a constant shearing stress. A platinum ring suspended from a torsion wire is kept under constant torsional stress and its rotation in the plane of the interface is measured as a function of time.

The study of surface rheology is useful in connection with the stability of emulsions and foams (Chapter 10) and the effectiveness of lubricants, adhesives, etc.

4.   *Electron micrographs of monolayers*—A technique for studying monolayers by electron microscopy has been developed recently.[60]

The films are transferred from the substrate on to a collodion support and shadow-cast by a beam of metal atoms directed at an angle α (about 15°) to the surface (Figure 4.22). If the width x of the uncoated regions is measured, the thickness of the film, $x \tan \alpha$, can be calculated; for example, a $n\text{-}C_{36}H_{73}COOH$ film has been

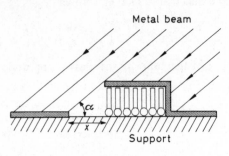

Figure 4.22

shown to be about 5 nm thick, i.e. consistent with a vertically orientated monomolecular layer. The technique has also been used for following the state of the surface as a film is compressed.

The physical states of monomolecular films

Two-dimensional monolayers can exist in different physical states which bear some resemblance to the solid, liquid and gaseous states in three-dimensional matter. Surface films are best classified according to the lateral adhesion between the film molecules, including end-groups. Factors such as ionisation (and hence the pH of the substrate) and temperature play an important part in determining the nature of the film. Monolayers can be roughly classified as:

1.   *Condensed* (solid) films in which the molecules are closely packed and steeply orientated toward the surface.
2.   Films which are still coherent but occupy a much larger area than condensed films. They have no real three-dimensional equivalent since they act as highly compressible liquids. A number of distinct types of these *expanded* films have been recognized,[10] the most important being the *liquid-expanded state*, but these will not be considered in detail.
3.   *Gaseous* or *vapour* films in which the molecules are separate

and move about the surface independently, the surface pressure being exerted on the barriers containing the film by a series of collisions.

Gaseous films

The principal requirements for an ideal gaseous film are that the constituent molecules must be of negligible size with no lateral adhesion between them. Such a film would obey an ideal two-dimensional gas equation, $\pi A = kT$, i.e. the $\pi - A$ curve would be a rectangular hyperbola. This ideal state of affairs is, of course, unrealisable but is approximated to by a number of insoluble films, especially at high areas and low surface pressures. Monolayers of soluble material are normally gaseous. If a surfactant solution is sufficiently dilute to allow solute–solute interactions at the surface to be neglected, the lowering of surface tension will be approximately linear with concentration, i.e.

$$\gamma = \gamma_0 - bc \text{ (where } b \text{ is a constant)}$$

Therefore $\qquad \pi = bc \quad$ and $\quad \mathrm{d}\gamma/\mathrm{d}c = -b$
Substituting in the Gibbs equation

$$\left(\Gamma = \frac{-c}{kT} \frac{\mathrm{d}\gamma}{\mathrm{d}c} \right)$$

gives $\qquad\qquad \Gamma = \frac{1}{A} = \frac{\pi}{kT}$

i.e. $\qquad\qquad\qquad \pi A = kT \qquad\qquad \dots \dots (4.26)$

where A is the average area per molecule.

An example of a gaseous film is that of cetyl trimethyl ammonium bromide (Figure 4.23). The molecules in the film are ionised to $C_{16}H_{33}N(CH_3)_3^+$ and, therefore, repel one another in the aqueous phase, so that π is relatively large at all points on the $\pi - A$ curve. Film pressures are greater for a given area at the oil–water interface than at the air–water interface, because the oil penetrates between the hydrocarbon chains of the film molecules and removes most of the inter-chain attraction. The $\pi - A$ curve at the air–water interface approximates to $\pi A = kT$, presumably because inter-chain attractions and electrical repulsions are of the same order of magnitude, whereas at the oil–water interface $\pi A > kT$, because the electrical repulsion between the film molecules outweighs the inter-chain attraction. Fatty acids and alcohols of chain-length C_{12} and less give imperfect gaseous films when spread on water at room

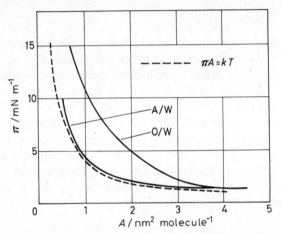

Figure 4.23. $\pi - A$ curves for cetyl trimethyl ammonium bromide at air–water and oil–water interfaces at 20°C

temperature, for which $\pi A < kT$, especially at high pressures and low areas.[2]

Condensed films

Palmitic, stearic and higher straight-chain fatty acids are examples

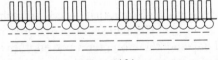

Figure 4.24

of materials which give condensed films at room temperature. At high film areas the molecules of fatty acid do not separate completely from one another, as the cohesion between the hydrocarbon chains is strong enough to maintain the film molecules in small clusters or islands on the surface (Figure 4.24). Because of this strong cohering tendency the surface pressure remains very low as the film is compressed and then rises rapidly when the molecules become tightly packed together.

For stearic acid spread on dilute HCl an initial pressure rise is observed at about $0.25 \ \text{nm}^2$ molecule^{-1} corresponding to the initial packing of the end-groups (Figure 4.25). The $\pi - A$ curve becomes

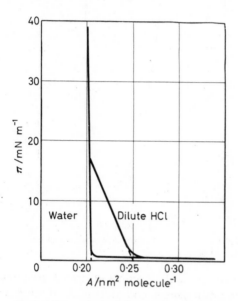

Figure 4.25. $\pi - A$ curves for stearic acid spread on water and dilute acid substrates at 20°C

very steep at about $0.205 \ \text{nm}^2$ molecule^{-1}, when it is supposed that more efficient packing has been achieved by staggering of the end-groups and interlocking of the hydrocarbon chains. A limiting area of $0.205 \ \text{nm}^2$ molecule^{-1} is observed for straight-chain fatty acids irrespective of the chain length. The packing of the molecules in the film at this point is not far short of that in the crystalline state. The cross-sectional area of stearic acid molecules from x-ray diffraction measurements is about $0.185 \ \text{nm}^2$ at normal temperatures. Any attempt to compress a condensed film beyond its limiting area will eventually lead to a collapse or buckling of the film.

Expanded films

Oleic acid (Figure 4.26) gives a much more expanded film than the corresponding saturated acid, stearic acid, i.e. π is greater for any value of A. Because of the double bond there is less cohesion between the hydrocarbon chains than for stearic acid. Also, at large

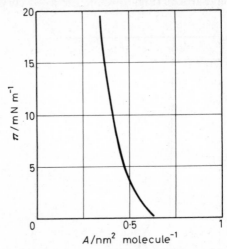

Figure 4.26. $\pi - A$ curve for oleic acid spread on water at 20°C

areas the oleic acid molecules become separated from one another, since the hydrocarbon chains tend to lie flat and independently on the surface with the hydrophilic double bonds in contact with the water. Compression of the oleic acid film forces the double bonds

Figure 4.27

above the surface and eventually orientates the hydrocarbon chains in a vertical position (Figure 4.27). This process is somewhat gradual as indicated by the form of the $\pi - A$ curve. In conformity with this, the rate of oxidation of an oleic acid monolayer by a dilute acid permanganate substrate is found to be greater at high areas.

There are a number of instances in which (with the aid of sensitive measurements) well defined transitions between gaseous and coherent states are observed as the film is compressed. The $\pi - A$ curves show a marked resemblance to Andrews' $p - V$ curves for the three-dimensional condensation of vapours to liquids. The $\pi - A$ curve for myristic acid, given as an example, has been drawn schematically to accentuate its main features (Figure 4.28). Above

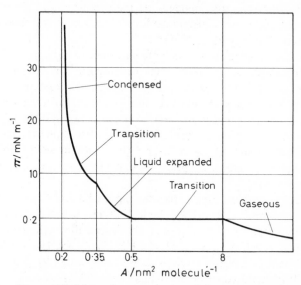

Figure 4.28. Schematic representation of the $\pi - A$ curve for myristic acid spread on 0·1 mol dm^{-3} HCl at 14°C

8 nm^2 molecule^{-1} the film is gaseous and a liquid-expanded film is obtained on compression to 0·5 nm^2 molecule^{-1}. Fluctuating surface film potentials verify the heterogeneous, transitional nature of the surface between 0·5 nm^2 molecule^{-1} and 8 nm^2 molecule^{-1}.

Liquid-expanded films obey the equation of state

$$(\pi - \pi_0)(A - A_0) = kT \qquad \qquad \dots . (4.27)$$

which resembles the van der Waals equation. The accepted theory of the liquid-expanded state, suggested by Langmuir, is that the monolayer behaves as a duplex film in which the head groups are in a state of two-dimensional kinetic agitation while the attractive forces between the hydrocarbon chains keep the film coherent.

Factors influencing the physical state of monomolecular films

As remarked previously, the physical state of a monolayer depends on the lateral cohesive forces between the constituent molecules. By a suitable choice of chain length and temperature, straight-chain fatty acids and alcohols, etc., can be made to exhibit the various monolayer states, one CH_2 group being equivalent to a temperature change of *ca.* 5K to 8K.

Lateral cohesion also depends on the geometry and orientation of the film molecules, so that the following factors will favour the formation of an expanded film:

1. Bulky head groups, which prevent efficient packing and hence maximum cohesion between the hydrocarbon chains.

2. More than one polar group; e.g. unsaturated fatty acids, hydroxy-acids. A film pressure is required to overcome the attraction between the second polar group and the aqueous substrate before the molecules can be orientated vertically.

3. More than one hydrocarbon chain orientated in different directions from the polar part of the molecule; e.g. esters, glycerides.

4. Bent hydrocarbon chains; e.g. brassidic acid (*trans*-$CH_3(CH_2)_7CH = CH(CH_2)_{11}COOH$), which has a straight hydrocarbon chain, gives a condensed film, whereas erucic acid (*cis*-$CH_3(CH_2)_7CH = CH(CH_2)_{11}COOH$), which has a bent hydrocarbon chain, gives a very expanded film.[61]

5. Branched hydrocarbon chains.

The nature of the substrate, particularly pH, is important when the monolayer is ionisable. When spread on alkaline substrates, because of the ionisation and consequent repulsion between the carboxyl groups, fatty acid monolayers form gaseous or liquid-expanded films at much lower temperatures. Dissolved electrolytes in the substrate can also have a profound effect on the state of the film; for example, Ca^{2+} ions form insoluble calcium soaps with fatty acid films (unless the pH is very low), thus making the film more condensed.

Evaporation through monolayers

1. *Water conservation*—The annual loss of water from hot country lakes and reservoirs due to evaporation is usually about 3 m per

year. This evaporation can be considerably reduced by coating the water surface with an insoluble monolayer; for example, a mono-layer of cetyl alcohol can reduce the rate of evaporation by as much as 40 per cent. Insoluble monolayers also have the effect of damping out surface ripples.

To attain minimum permeability to evaporation, a close-packed monolayer under a sufficient state of compression to squeeze out any surface impurities is required. The monolayer must also be self-healing in response to adverse meteorological conditions such as wind, dust and rainfall, so that spreading ability is also needed. To compromise between these requirements, commercial cetyl alcohol (which also contains some steryl, myristyl and oleyl alcohol) has been used successfully.

The monolayer can be spread initially either from solvent or as a powder, the latter being preferred. Small rafts which allow mono-layer material to seep out slowly and replace losses are usually installed at points on the water surface. Oxygen diffuses readily through insoluble monolayers. The oxygen content of the under-lying water is somewhat less (80 per cent saturated rather than the normal 90 per cent saturated) since the surface is more quiescent, but this has no adverse effect on life beneath the surface.

2. *Monolayers on droplets*—Evaporation from small water droplets used to bind fine dust in coal mines is severe unless the surfaces of the droplets are covered with a protective insoluble film. By previously dispersing a little cetyl alcohol in the water the life of the droplets can be increased some thousand-fold. Tarry material, dust, etc., in town fogs is responsible for delay in dispersal because of monolayer formation.

Surface films of proteins

Surface films of high-polymer material, particularly proteins, offer another wide field of study.

Proteins consist of a primary structure of amino acid residues connected in a definite sequence by peptide linkages to form poly-peptide chains

$$-\underset{\underset{R}{|}}{CH}-NH-CO-\underset{\underset{R'}{|}}{CH}-NH-CO-\underset{\underset{R''}{|}}{CH}-\ etc.$$

which may embrace hundreds of such residues. These polypeptide chains normally take up a helical configuration, which is stabilised mainly by hydrogen bonding between spatially adjacent —NH— and —CO— groups. The helical polypeptide chains of the globular proteins are in turn folded up to give compact and often nearly spherical molecules. This configuration is maintained by hydrogen bonding, van der Waals forces between the non-polar parts, di-sulphide cross-linking, etc.

Any significant alteration in this arrangement of the polypeptide chains without damage to the primary structure is termed *denaturation*. The common agents of denaturation are those which would be expected to modify hydrogen bonds and other weak stabilising linkages, e.g. acids, alkalis, alcohol, urea, heat, ultraviolet light and surface forces. Protein denaturation is accompanied by a marked loss of solubility and is usually, but not necessarily, an irreversible process. Proteins adsorb and denature at high-energy air–water and oil–water interfaces because unfolding allows the polypeptide chains to be orientated with most of their hydrophilic groups in the water phase and most of the hydrophobic groups away from the water phase.

If a small amount of protein solution is suitably spread at the surface of an aqueous substrate, most of the protein will be surface denatured, giving an insoluble monomolecular film before it has a chance to dissolve. The techniques already described for studying spread monolayers of insoluble material can, therefore, be used in the study of protein films. One frequently-used spreading solution contains about 0·1 per cent protein in a mixture of alcohol and aqueous sodium acetate.

Compression of protein films below about $1 \, \text{m}^2 \, \text{mg}^{-1}$ results in close packing of the polypeptide chains and the gradual development of a gel-like structure. At a surface pressure of *ca.* 15–20 mN m^{-1} a time-dependent collapse of the film into bundles of insoluble fibres takes place. The area occupied by the compressed protein film is often characterised by the limiting area obtained by extrapolating the approximately linear part of the $\pi - A$ curve to zero pressure. A more satisfactory characterisation of the close-packed film, especially in relation to the areas indicated by x-ray diffraction measurements on protein fibres, is given by the area at minimum compressibility,[62] i.e. where the $\pi - A$ curve is at its steepest. This area corresponds approximately to the onset of film collapse. Not surprisingly, these areas show little variation from protein to protein, since the average size of the constituent amino acid residues does not alter greatly.

Many of the proteins from which these films can be formed are approximately spherical in the native state, with diameters of *ca.* 5 nm to 10 nm. Since a limiting area of 1 m² mg⁻¹ corresponds to about 0·15 nm² per peptide residue, or a film only 0·8 nm to 1·0 nm

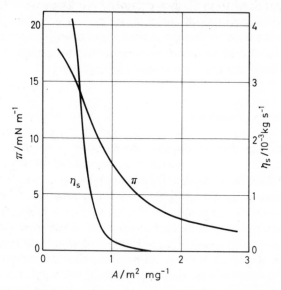

Figure 4.29. $\pi - A$ and $\eta_s - A$ curves for a spread monolayer of β-globulin at the petrol ether–water interface[63] (By courtesy of The Faraday Society)

thick, then clearly some unfolding of the polypeptide chains takes place at the surface. Proteins unfold even further at oil–water interfaces.

At low compressions, up to *ca.* 1 mN m⁻¹, protein films tend to be gaseous, thus permitting relative molecular mass determination.

For an ideal gaseous film

$$\pi A_m = RT$$

where A_m is the molar area of the film material. Therefore,

$$\pi AM = RT$$

where A represents the area per unit mass and M is the molar mass of the film material. To realise ideal gaseous behaviour experimental data must be extrapolated to zero surface pressure, i.e.

$$\lim_{\pi \to 0} \pi A = \frac{RT}{M} \qquad \qquad \ldots . (4.28)$$

The relative molecular masses of a number of spread proteins have been determined from surface-pressure measurements.[64] In many cases they compare with the relative molecular masses in bulk solution. Relative molecular masses significantly lower than the bulk solution values have also been reported, which suggests surface dissociation of the protein molecules into sub-units.

Interactions in mixed films

Mixed surface films, especially those likely to be of biological importance, have been subjected to a great deal of investigation.

Often there is evidence of interaction between stoichiometric proportions of the components of a mixed monolayer. Evidence of interaction can be sought by measuring partial molecular areas or by studying the collapse of the mixed film. The partial molecular areas of the components of a mixed film are usually different from the molecular areas of the pure components when interaction occurs. A mixed monolayer may collapse in one of two ways: (1) with no interaction one component displaces the other, usually at the collapse pressure of the displaced material; or (2) interacting mixed films collapse as a whole at a surface pressure different from, and usually greater than, the collapse pressure of either component.

Another type of interaction is the penetration of a surface-active constituent of the substrate into a spread monolayer. Penetration effects can be studied by injecting a solution of the surface-active material into the substrate immediately beneath the monolayer: (1) if there is no association between the injected material and the monolayer, π and ΔV will both remain unaltered; (2) if the injected material adsorbs on to the underside of the monolayer without actual penetration, ΔV will change appreciably but π will alter very little; (3) if the injected material penetrates into the monolayer (i.e. when there is association between both polar and non-polar parts of the injected and original monolayer materials), π will change significantly and ΔV will assume an intermediate value between ΔV of the original monolayer and ΔV of a monolayer of injected material. Penetration is less likely to occur when the monolayer is tightly packed.

Natural membranes consist mainly of lipo-protein material, and artificially prepared monolayers have been successfully used as models for studying certain biological processes. Schulman and Rideal[65] investigated the action of agglutinating (coagulating) and lytic agents on red blood cells by the monolayer technique. A mixed film of 20 per cent cholesterol and 80 per cent gliadin was used to represent the red blood cell membrane. It was found that lytic agents penetrate into this monolayer, whereas agglutinating agents only adsorb on to the underside of the film. This implies that red blood cells are altered structurally (lysis) due to the penetration of lytic agents into the membrane, whilst agglutinating agents merely adsorb on to the membrane surface.

5
The Solid–Gas Interface

ADSORPTION OF GASES AND VAPOURS ON SOLIDS

When a gas or vapour is brought into contact with a clean solid surface some of it will become attached to the surface in the form of an adsorbed layer. The solid is generally referred to as the *adsorbent,* and the gas or vapour as the *adsorbate.* It is possible that uniform absorption into the bulk of the solid might also take place, and, since *adsorption* and *absorption* cannot always be distinguished experimentally, the generic term *sorption* is sometimes used to describe the general phenomenon of gas uptake by solids.

Any solid is capable of adsorbing a certain amount of gas, the extent of adsorption at equilibrium depending on temperature, the pressure of the gas and the effective surface area of the solid. The most notable adsorbents are, therefore, highly porous solids, such as charcoal and silica gel (which have large internal surface areas—up to *ca.* $1\,000 \text{ m}^2 \text{ g}^{-1}$) and finely divided powders. The relationship at a given temperature between the equilibrium amount of gas adsorbed and the pressure of the gas is known as the adsorption isotherm (Figures, 5.1, 5.6, 5.7, 5.8, 5.10, 5.12).

Adsorption reduces the imbalance of attractive forces which exists at a surface, and hence the surface free energy of a heterogeneous system. In this respect, the energy considerations relating to solid surfaces are, in principle, the same as those already discussed for liquid surfaces. The main differences between solid and liquid surfaces arise from the fact that solid surfaces are heterogeneous in respect of activity, with properties dependent, to some extent, on previous environment.

Physical adsorption[9, 66, 67] and chemisorption[68, 69]

The forces involved in the adsorption of gases and vapours by solids may be: non-specific (van der Waals) forces, similar to the forces

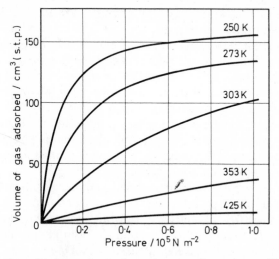

Figure 5.1. Adsorption isotherms for ammonia on charcoal[70]

involved in liquefaction; or stronger specific forces, such as those which are operative in the formation of chemical bonds. The former are responsible for *physical adsorption* and the latter for *chemisorption*.

When adsorption takes place, the gas molecules are restricted to two-dimensional motion. Gas adsorption processes are, therefore, accompanied by a decrease in entropy. Since adsorption also involves a decrease in free energy, then from the thermodynamic relationship

$$\Delta G = \Delta H - T\Delta S \qquad \ldots . (5.1)$$

it is evident that $\Delta H_{ads.}$ must be negative, i.e. the adsorption of gases and vapours on solids is always an exothermic process.* The extent of gas adsorption (under equilibrium conditions), therefore, increases with decreasing temperature (see Figure 5.1). Heats of adsorption can be measured by direct calorimetric methods. Isosteric (constant adsorption) heats of adsorption can be calculated from reversible adsorption isotherms using the Clausius-Clapeyron equation

$$\left(\frac{\partial \ln p}{\partial T} \right)_V = \frac{-\Delta H_{ads.}}{RT^2} \qquad \dots \dots (5.2)$$

The heats of physical adsorption of gases are usually similar to their heats of condensation; for example, the integral heat of physical adsorption of nitrogen on an iron surface is $ca. -10$ kJ mol^{-1} (the heat of liquefaction of nitrogen is -5.7 kJ mol^{-1}). Heats of chemisorption are generally much larger; for example, the integral heat of chemisorption of nitrogen on iron is $ca. -150$ kJ mol^{-1}.

The attainment of physical adsorption equilibrium is usually rapid, since there is no activation energy involved, and (apart from complications introduced by capillary condensation) the process is readily reversible. Multilayer physical adsorption is possible, and at the saturated vapour pressure of the gas in question physical adsorption becomes continuous with liquefaction.

Only monomolecular chemisorbed layers are possible. Chemisorption is a specific process which may require an activation energy and may, therefore, be relatively slow and not readily reversible. The nature of physical adsorption and chemisorption is illustrated by the schematic potential energy curves shown in Figure 5.2 for the adsorption of a diatomic gas X_2 on a metal M.

Curve P represents the physical interaction energy between M and X_2. It inevitably includes a short range negative (attractive) contribution arising from London-van der Waals dispersion forces and an even shorter range positive contribution (Born repulsion) due to an overlapping of electron clouds. It will also include a further van der Waals attractive contribution if permanent dipoles are involved. The nature of van der Waals forces is discussed on page 169.

Curve C represents chemisorption in which the adsorbate X_2 dissociates to $2X$. For this reason, an energy equal to the dissociation

---

* This argument does not necessarily hold for processes such as adsorption from solution and micellisation, since a certain amount of destructuring (e.g. desolvation) may be involved and the net entropy change may be positive.

energy of X_2 is represented at large distances. The curve is also characterised by a relatively deep minimum which represents the heat of chemisorption, and which is at a shorter distance from the solid surface than the relatively shallow minimum in the physical adsorption curve.

It can be seen from these curves that initial physical adsorption is a most important feature of chemisorption. If physical adsorption

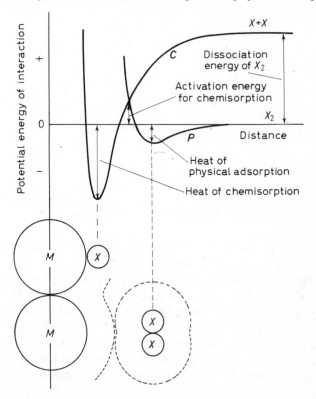

Figure 5.2. Potential energy curves for physical adsorption and chemisorption

were non-existent the energy of activation for chemisorption would be equal to the high dissociation energy of the adsorbate gas molecules. As it is, an adsorbate gas molecule is first physically adsorbed, which involves approaching the solid surface along a low energy path. Transition from physical adsorption to chemisorption takes place at the point where curves P and C intersect and the energy at this point is equal to the activation energy for chemisorption. The magnitude of this activation energy depends, therefore, on the shapes of the physical adsorption and chemisorption curves, and varies widely from system to system; for example, it is low for the chemisorption of hydrogen on to most metal surfaces.

If the activation energy for chemisorption is appreciable, the rate

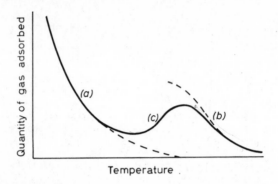

Figure 5.3. Schematic adsorption isobar showing the transition between physical adsorption and chemisorption

of chemisorption at low temperatures may be so slow that, in practice, only physical adsorption is observed.

Figure 5.3 shows how the extent of gas adsorption on to a solid surface might vary with temperature at a given pressure. Curve (*a*) represents physical adsorption equilibrium and curve (*b*) represents chemisorption equilibrium. The extent of adsorption at temperatures where the rate of chemisorption is slow, but not negligible, is represented by a non-equilibrium curve, such as (*c*), the location of which depends on the time allowed for equilibration.

Experimental methods for studying gas adsorption

The solid adsorbent under investigation must first be freed, as far as possible, from previously adsorbed gases and vapours. Evacuation to *ca.* 10^{-4} Torr (outgassing) for several hours will generally remove physically adsorbed gas. It is difficult, and often impossible, to remove chemisorbed gas completely unless the solid is heated to a high temperature. Such treatment might alter the sorptive capacity of the solid.

The adsorption of a gas or vapour is measured by admitting a known amount of the adsorbate into the evacuated, leak-free space containing the outgassed adsorbent. The extent of adsorption can then be determined either volumetrically or gravimetrically.

The volumetric method is mainly used for the purpose of determining specific surface areas of solids from gas (particularly nitrogen) adsorption measurements (see page 115). The gas is contained in a gas burette, and its pressure is measured with a manometer (see Figure 5.4). All of the volumes in the apparatus are calibrated so that when the gas is admitted to the adsorbent sample the amount adsorbed can be calculated from the equilibrium pressure reading. The adsorption isotherm is obtained from a series of measurements at different pressures.

The gravimetric method can be used for studying both gas and

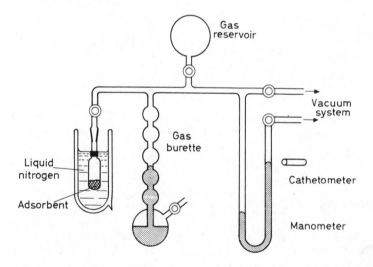

Figure 5.4. Volumetric apparatus for measuring gas adsorption at 77 K

vapour sorption. The outgassed sorbent is contained in a small bucket suspended from a quartz spiral which has previously been calibrated (see Figure 5.5). Purified gas or vapour is introduced into the evacuated apparatus, the pressure is noted and the extent of sorption is measured directly (allowing for buoyancy) as the increase in the weight of the sorbent sample. Again, an adsorption isotherm can be obtained from measurements at a given temperature covering a range of pressures.

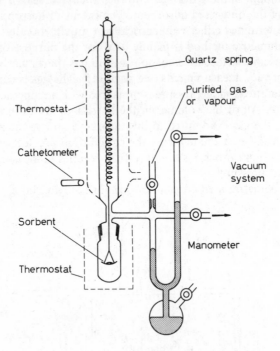

Figure 5.5. McBain–Bakr sorption balance

Classification of adsorption isotherms

Three phenomena may be involved in physical adsorption:
1. Monomolecular adsorption.
2. Multimolecular adsorption.
3. Condensation in pores or capillaries.

Frequently, there is overlapping of these phenomena, and the interpretation of adsorption studies can be complicated. Brunauer[71] has classified adsorption isotherms into the five characteristic types shown in Figure 5.6.

Type I isotherms (e.g. ammonia on charcoal at 273K) show a

fairly rapid rise in the amount of adsorption with increasing pressure up to a limiting value. They are referred to as Langmuir-type isotherms and are obtained when adsorption is restricted to a monolayer. Chemisorption isotherms, therefore, approximate to this shape. Type I isotherms have also been found for physical adsorption on solids containing a very fine pore structure.

Type II isotherms (e.g. nitrogen on silica gel at 77K) are frequently encountered, and represent multilayer physical adsorption on non-porous solids. They are often referred to as sigmoid isotherms. For

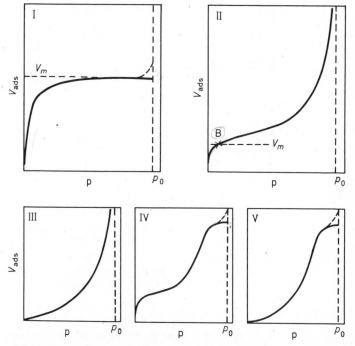

Figure 5.6. Brunauer's classification of adsorption isotherms[71] (p_0 = saturated vapour pressure)

such solids, point B represents the formation of an adsorbed monolayer. Physical adsorption on microporous solids can also result in type II isotherms. In this case, point B represents both monolayer adsorption on the surface as a whole and condensation in the fine pores. The remainder of the curve represents multilayer adsorption as for non-porous solids.

Type IV isotherms (e.g. benzene on ferric oxide gel at 320K) level off near the saturation vapour pressure and are considered to reflect capillary condensation in porous solids, the effective pore diameters generally being between about 2 nm and 20 nm. The upper limit of adsorption is mainly governed by the total pore volume.

Types III (e.g. bromine on silica gel at 352K) and V (e.g. water

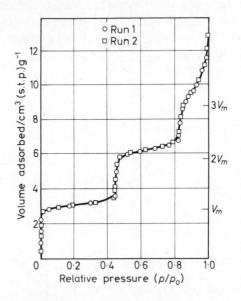

Figure 5.7. Stepwise isotherm for the adsorption of krypton at 90K on carbon black (graphitised at 3 000 K)[72] (By courtesy of *The Canadian Journal of Chemistry*)

vapour on charcoal at 373K) show no rapid initial uptake of gas, and occur when the forces of adsorption in the first monolayer are relatively small. These isotherms are rare.

Many adsorption isotherms are borderline cases between two or more of the above types. In addition, there are some isotherms which do not fit into Brunauer's classification at all, the most notable being the stepwise isotherms, an example of which is given in Figure 5.7. Stepwise isotherms are generally associated with adsorption on to uniform solid surfaces, each step corresponding to the formation of a complete monomolecular adsorbed layer (see page 113).

Capillary condensation theory

It has previously been shown (page 57) that the vapour pressure over a convex liquid surface is greater than that over the corresponding flat surface. A liquid which wets the wall of a capillary will have a concave liquid–vapour interface and, therefore, a lower vapour pressure in the capillary than in the bulk phase. This vapour pressure difference is given by the Kelvin equation, written in the form

$$RT \ln p_r/p_0 = -\frac{2\gamma V_m \cos \theta}{r} \qquad \dots (5.3)$$

where r is the radius of the capillary, and θ the contact angle between the liquid and the capillary wall.

Condensation can, therefore, take place in narrow capillaries at pressures which are lower than the normal saturation vapour pressure. Zsigmondy (1911) suggested that this phenomenon might also apply to porous solids. Capillary rise in the pores of a solid will usually be so large that the pores will tend to be either completely full of capillary condensed liquid or completely empty. Ideally, at a certain pressure below the normal condensation pressure all the pores of a certain size and below will be filled with liquid and the rest will be empty. It is probably more realistic to assume that an adsorbed monomolecular film exists on the pore walls before capillary condensation takes place. By a corresponding modification of the pore diameter, an estimate of pore size distribution (which will only be of statistical significance because of the complex shape of the pores) can be obtained from the adsorption isotherm.

Capillary condensation cannot account for multilayer adsorption at planar or convex surfaces, so that, although capillary condensation is undoubtedly a most important feature of physical adsorption on porous solids, it does not give the complete picture.

The capillary condensation theory provides a satisfactory explanation of the phenomenon of adsorption hysteresis, which is frequently observed for porous solids. Adsorption hysteresis is a term which is used when the desorption isotherm curve does not coincide with the adsorption isotherm curve (Figure 5.8).

A possible explanation of this phenomenon is given in terms of contact angle hysteresis. The contact angle on adsorption, when liquid is advancing over a dry surface, is generally greater than the contact angle during desorption, when liquid is receding from a wet surface. From the Kelvin equation, it is evident that the pressure below which liquid vaporises from a particular capillary will, under these circumstances, be lower than the pressure required for capillary condensation.

Another theory of adsorption hysteresis considers that there are two types of pores present, each having a size distribution. The first type are V-shaped, and these fill and empty reversibly. The second type have a narrow neck and a relatively wide interior. These 'ink-bottle' pores are supposed to fill completely when a p/p_0 value

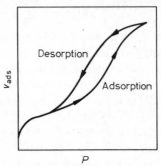

Figure 5.8. A hysteresis loop in physical adsorption

corresponding to the relatively wide pore interior is reached, but once filled they retain their contents until p/p_0 is reduced to a value corresponding to the relatively small width of the pore neck.

Isotherm equations

Numerous attempts have been made at developing mathematical expressions from postulated adsorption mechanisms to fit the various experimental isotherm curves. The three isotherm equations which are most frequently used are those due to Langmuir, to Freundlich, and to Brunauer, Emmett and Teller (BET).

1.   *The Langmuir Adsorption Isotherm*—Before 1916, adsorption theories postulated either a condensed liquid film or a compressed gaseous layer which decreases in density as the distance from the surface increases. Langmuir (1916) was of the opinion that, because of the rapidity with which intermolecular forces fall off with distance, adsorbed layers are not likely to be more than one molecular

layer in thickness. This view is generally accepted for chemisorption and for physical adsorption at low pressures and moderately high temperatures.

The Langmuir adsorption isotherm is based on the characteristic assumptions that (a) only monomolecular adsorption takes place, (b) adsorption is localised, and (c) the heat of adsorption is independent of surface coverage. A kinetic derivation follows in which the velocities of adsorption and desorption are equated with one another to give an expression representing adsorption equilibrium.

Let V equal the equilibrium volume of gas adsorbed per unit mass of adsorbent at a pressure p and V_m equal the volume of gas required to cover unit mass of adsorbent with a complete monolayer.

The velocity of adsorption depends on: (a) the rate at which gas molecules collide with the solid surface, which is proportional to the pressure; (b) the probability of striking a vacant site $(1 - V/V_m)$; and (c) an activation term $\exp[-E/RT]$, where E is the activation energy for adsorption.

The velocity of desorption depends on: (a) the fraction of the surface which is covered, V/V_m; and (b) an activation term $\exp[-E'/RT]$, where E' is the activation energy for desorption.

Therefore, when adsorption equilibrium is established

$$p \, (1 - V/V_m) \exp[-E/RT] = k \, (V/V_m) \exp[-E'/RT]$$

where k is a proportionality constant, i.e.

$$p = k \exp[\Delta H_{ads.}/RT] \frac{V/V_m}{(1 - V/V_m)}$$

where $\Delta H_{ads.} = E - E' =$ heat of adsorption (negative).

Assuming that the heat of adsorption, $\Delta H_{ads.}$ is independent of surface coverage

$$k \exp[\Delta H_{ads.}/RT] = 1/a$$

where a is a constant dependent on the temperature, but independent of surface coverage. Therefore

$$ap = \frac{V/V_m}{(1 - V/V_m)} \qquad \dots (5.4)$$

or

$$V = \frac{V_m ap}{(1 + ap)} \qquad \dots (5.5)$$

$$= \frac{V}{V_m} = \theta = \frac{ap}{1 + ap} \qquad a =$$

or $$p/V = p/V_m + 1/aV_m \qquad \qquad \dots (5.6)$$

i.e. a plot of p/V versus p should give a straight line of slope $1/V_m$ and an intercept of $1/aV_m$ on the p/V axis.

At low pressures the Langmuir isotherm equation reduces to $V = V_m ap$, i.e. the volume of gas adsorbed varies linearly with pressure. At high pressures a limiting monolayer coverage, $V = V_m$, is reached. The curvature of the isotherm at intermediate pressures depends on the value of the constant a, and hence on the temperature.

The most notable criticism of the Langmuir adsorption equation

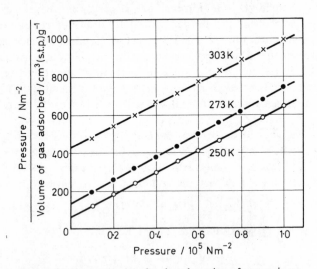

Figure 5.9. Langmuir plots for the adsorption of ammonia on charcoal shown in *Figure 5.1*. Slope $= 1/V_m$

concerns the simplifying assumption that the heat of adsorption is independent of surface coverage, which, as discussed in the next section, is not likely to be the case. Nevertheless, many experimental adsorption isotherms fit the Langmuir equation reasonably well.

When the components of a gas mixture compete for the adsorption sites on a solid surface the Langmuir equation takes the general form

$$\frac{V_i}{V_{m,i}} = \frac{a_i p_i}{1 + \Sigma a_j p_j} \qquad \qquad \dots (5.7)$$

2.   *The Freundlich (or Classical) Adsorption Isotherm*—The variation of adsorption with pressure can often be represented (especially at moderately low pressures) by the equation

$$V = kp^{1/n} \qquad \ldots (5.8)$$

where k and n are constants, n generally being greater than unity. Taking logarithms

$$\log V = \log k + 1/n \log p \qquad \ldots (5.9)$$

i.e. a plot of $\log V$ versus $\log p$ should give a straight line.

This adsorption equation was originally proposed on a purely empirical basis. It can be derived theoretically, however, for an adsorption model in which the magnitude of the heat of adsorption varies exponentially with surface coverage. The Freundlich equation is, in effect, the summation of a distribution of Langmuir equations; however, the volume of gas adsorbed is not depicted as approaching a limiting value as in a single Langmuir equation.

3.   *The BET equation for multimolecular adsorption*—Because the forces acting in physical adsorption are similar to those operating in liquefaction (i.e. van de Waals forces), physical adsorption (even on flat and convex surfaces) is not limited to a monomolecular layer, but can continue until a multimolecular layer of liquid covers the adsorbent surface.

The theory of Brunauer, Emmett and Teller[73] is an extension of the Langmuir treatment to allow for multilayer adsorption on non-porous solid surfaces. The BET equation is derived by balancing the rates of evaporation and condensation for the various adsorbed molecular layers, and is based on the simplifying assumption that a characteristic heat of adsorption ΔH_1 applies to the first mono-layer, whilst the heat of liquefaction ΔH_L of the vapour in question applies to adsorption in the second and subsequent molecular layers. The equation is usually written in the form

$$\frac{p}{V(p_0 - p)} = \frac{1}{V_m c} + \frac{(c-1)}{V_m c}\frac{p}{p_0} \qquad \ldots (5.10)$$

where p_0 is the saturation vapour pressure, V_m the monolayer capacity, and $c \approx \exp[(\Delta H_L - \Delta H_1)/RT]$.

The main purpose of the BET equation is to describe type II isotherms. In addition, it reduces to the Langmuir equation at low pressures; and type III isotherms are given in the unusual circumstances when monolayer adsorption is less exothermic than liquefaction, i.e. $c < 1$ (see Figure 5.10).

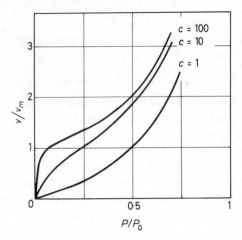

Figure 5.10. BET isotherms

The BET model can also be applied to a situation which might be applicable to porous solids. If adsorption is limited to n molecular layers (where n is related to the pore size), the equation

$$V = \frac{V_m c x}{(1-x)} \cdot \frac{1-(n+1)x^n + n x^{n+1}}{1+(c-1)x-c x^{n+1}} \quad \dots (5.11)$$

is obtained, where $x = p/p_0$. This equation is, in fact, a general expression which reduces to the Langmuir equation when $n = 1$ and to the BET equation when $n = \infty$.

Adsorption energies

A most important feature of the models upon which adsorption isotherm equations, such as those above, are based is a characteristic assumption relating to heat of adsorption and surface coverage. Several factors merit consideration in this respect.

Solid surfaces are usually heterogeneous; therefore, since adsorption at the more active sites is favoured, heats of both monolayer physical adsorption and chemisorption might, in this respect, be expected to become significantly less exothermic as the surface coverage increases, as, for example, shown at low pressures in Figures 5.11a and 5.11b. This, in turn, would cause the initial slope of an adsorption isotherm to be steeper than that predicted according to the Langmuir equation or the BET equation.

Chemisorption might involve the adsorbate gas molecules either giving up electrons to or receiving electrons from the adsorbent solid. As either of these processes continues, further adsorption becomes more and more difficult and monolayer coverage is not as readily attained as would be predicted according to the Langmuir

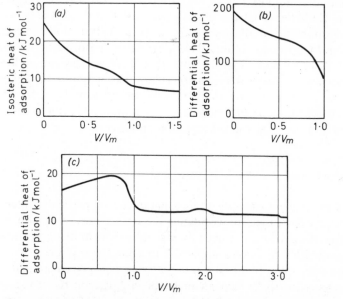

Figure 5.11. Adsorption energy and surface coverage. (a) Physical adsorption of nitrogen on rutile at 85 K.[74] (b) Chemisorption of hydrogen on tungsten.[75] (c) Physical adsorption of krypton on graphitised carbon black.[72] (See Figure 5.7.) (By courtesy of (a) Science Progress, (b) Discussions of the Faraday Society, and (c) The Canadian Journal of Chemistry)

equation. The heat of adsorption becomes less exothermic as monolayer coverage is approached, as, for example, shown in Figure 5.11b.

When a gas molecule is adsorbed on to a solid surface which is already partially covered with a monomolecular layer, lateral interaction with the adsorbed gas molecules will be involved in addition to interaction with the solid. In this respect, the heat of adsorption might be expected to become more exothermic with increasing surface coverage, as, for example, shown in Figure 5.11c.

The shape of a multilayer physical adsorption isotherm depends on the tendency of each adsorbed monomolecular layer (particularly the first layer) to be completed before any adsorption into further layers takes place. This situation is favoured if the adsorption energy for the layer being completed is significantly more exothermic than that for commencing further adsorbed layers. As a rather extreme example, Figure 5.11c shows this kind of variation of adsorption energy with surface coverage for the physical adsorption of a gas on a fairly homogeneous solid surface. The corresponding adsorption isotherm (Figure 5.7) shows at least two distinct steps each corresponding to the formation of an adsorbed monomolecular layer. In most cases of multilayer physical gas adsorption, however, the adsorption energies are such that there is a greater or lesser tendency for the first monomolecular adsorbed layer to be completed prior to any adsorption into the second monolayer, but little tendency for the second monolayer to be completed prior to adsorption into the third and subsequent monolayers.

Surface areas

The monolayer capacity V_m is a parameter of particular interest, since it can be used for calculating the surface area of an adsorbent if the effective area occupied by each adsorbate molecule is known.

If the BET equation is applicable to a multilayer physical adsorption isotherm, a plot of $p/V(p_0-p)$ versus p/p_0 gives a straight line of slope $(c-1)/V_m c$ and an intercept of $1/V_m c$ on the $p/V(p_0-p)$ axis, i.e.

$$V_m = \frac{1}{\text{Slope} + \text{Intercept}} \qquad \dots (5.12)$$

Using the appropriate gas equation, the monolayer capacity can be calculated in terms of adsorbed molecules per unit mass of adsorbent.

Although the BET equation is open to a great deal of criticism, because of the simplified adsorption model upon which it is based, it nevertheless fits many experimental multilayer adsorption isotherms particularly well at pressures between about $0.05 p_0$ and $0.35 p_0$ (within which range the monolayer capacity is generally reached). However, with porous solids (for which adsorption hysteresis is characteristic), or when point **B** on the isotherm (Figure 5.6) is not very well defined, the validity of values of V_m calculated using the BET equation is doubtful.

The adsorbate most commonly used for BET surface area determinations is nitrogen at 77K (liquid nitrogen temperature). The

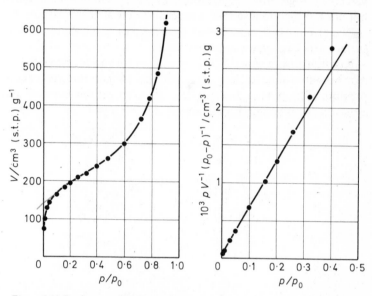

Figure 5.12. Isotherm and BET plot for the multilayer adsorption of nitrogen on a non-porous sample of silica gel at 77 K

effective area occupied by each adsorbed nitrogen molecule at monolayer capacity can be calculated from the density of liquid nitrogen (0·81 g cm^{-3}) on the basis of a model of close-packed spheres. The value so obtained is $16\cdot2 \times 10^{-20}$ m^2. Adoption of this value leads to BET areas for non-porous solids which are generally in accord with corresponding surface areas determined by other techniques.

With most other adsorbates, adjusted molecular areas (determined, for example, by calibration with nitrogen adsorption data) which are generally in excess of those calculated from liquid densities must be used, and, moreover, for a given adsorbate, this adjusted value usually varies from solid to solid. This is due mainly to a certain amount of localisation of the adsorbed gas in the first monomolecular layer with respect to the variously distributed potential energy minima of the solid lattices. To avoid excessive localisation a low value of $-\Delta H_1$ is desirable; however, a high value of c (i.e. $-\Delta H_1 \gg -\Delta H_L$) is also required to give a well defined point B. The main reason why nitrogen is a particularly suitable adsorbate for surface area determinations is because the value of c is generally high enough to give a well defined point B, but not too high to give excessive localisation of adsorption.

Effective cross-sectional areas of molecules adsorbed on to solid surfaces are catalogued in Reference 76.

Krypton adsorption at 77K is often used for the determination of relatively low solid surface areas. At this temperature the vapour pressure of krypton (and so the dead space correction) is small, and a reasonable precision is attainable.

6
The Solid–Liquid Interface

CONTACT ANGLES AND WETTING

When a drop of liquid is placed on a flat solid surface it may spread completely over the surface or, more likely, it may remain as a drop having a definite angle of contact with the solid surface (Figure 6.1).

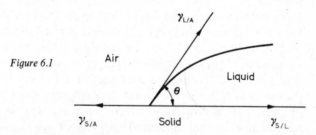

Figure 6.1

Air

Liquid

Solid

$\gamma_{L/A}$

$\gamma_{S/A}$

$\gamma_{S/L}$

θ

Assuming that the various surface forces can be represented by surface tensions acting in the direction of the surfaces, then, equating the horizontal components of these tensions

$$\gamma_{S/A} = \gamma_{S/L} + \gamma_{L/A} \cos \theta$$

Combining this expression with the appropriate form of the Dupré equation (4.20) (i.e. $W_{S/L} = \gamma_{S/A} + \gamma_{L/A} - \gamma_{S/L}$)

$$W_{S/L} = \gamma_{L/A} (1 + \cos \theta) \qquad \ldots \ldots (6.1)$$

117

which is known as Young's equation. Therefore, zero contact angle results when the forces of attraction between liquid and solid are equal to or greater than those between liquid and liquid, and a finite contact angle results when the liquid adheres to the solid less than it coheres to itself.

The solid is completely wetted by the liquid if the contact angle is zero and only partially wetted if the contact angle is finite. Complete non-wetting implies a contact angle of 180°, which is an unrealistic situation since it requires that either $W_{S/L} = 0$ or $\gamma_{L/A} = \infty$. There is always some solid–liquid attraction; for example, water droplets will adhere to the underside of a paraffin wax surface ($\theta \sim 110°$). Some writers correlate $\theta < 90°$ with wetting and $\theta > 90°$ with non-wetting in order to be in approximate keeping with visual appearances; this is not entirely satisfactory.

Measurement of contact angles

The measurement of contact angles is complicated by the following factors:

1. Contamination of the liquid surface tends to reduce the contact angle. This follows from Young's equation since $\gamma_{L/A}$ is lowered and $W_{S/L}$ remains constant, and θ is therefore lowered.

2. Solid surfaces differ from liquid surfaces in that they show a far greater degree of heterogeneity, even after careful polishing; for example, a solid surface polished to the best optical standards is wavy and pitted compared with a quiescent liquid surface. To obtain a solid surface free from the impurities which are likely to have a significant effect on its properties is usually very difficult. It can, therefore, be appreciated that any measured property of a solid surface is subject to variability as a result of unavoidable sample differences.

3. Contact angles are often not definite quantities but differ according to whether the liquid is advancing over a dry surface or receding from a wet surface. This hysteresis effect is most noticeable with impure surfaces; the difference between advancing and receding contact angles may be as much as 50°. A common example of contact angle hysteresis is given by rain drops on a dirty window pane.

The measurement of contact angles is facilitated if a moderately large and uniform solid surface is available. One method is to adjust the angle of a plate immersed in the liquid so that the liquid surface remains perfectly flat right up to the solid surface (Figure 6.2). Another method, originally devised by Langmuir and Schaeffer, is

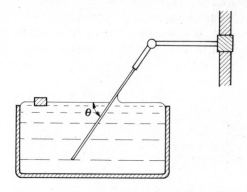

Figure 6.2. Tilting plate method for the measurement of contact angles

based on observation of the angle at which light from a point source is reflected from the surface of a liquid drop at its contact point with a plane solid surface.[77, 78] This technique has been refined for measuring contact angles formed by liquids on the surfaces of small-diameter filaments.[78]

The contact angles of solids in a finely divided form are technically important (e.g. flotation, page 122) but are difficult to measure. Bartell *et al.*[79] have developed a method based on displacement pressures for measuring such contact angles. The finely divided solid is packed into a tube and the resulting porous plug is considered to act as a bundle of capillaries of some average radius r. The pressure required to prevent the liquid in question from entering the capillaries of this porous plug is measured, and the general equation for capillary rise (4.3) is applied

$$p = 2\gamma \cos \theta/r \qquad \qquad \ldots (6.2)$$

The equivalent radius of the capillaries is then found from a similar experiment using a liquid which completely wets the solid, i.e.

$$p' = 2\gamma'/r \qquad \qquad \ldots (6.3)$$

Factors affecting contact angles

The contact angle between water and glass is increased considerably by even less than an adsorbed monolayer of greasy material such as fatty acid. $W_{S/L}$ is decreased since some of the glass–water interface is replaced by hydrocarbon–water interface (Figure 6.3a); hence, from Young's equation, θ increases.

The spreading of water on a hydrophobic solid surface is considerably helped by the addition of surface-active agents. $W_{S/L}$ is

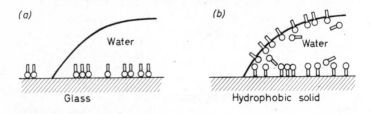

Figure 6.3

increased and $\gamma_{L/A}$ is decreased (Figure 6.3b) so that, from Young's equation, θ is reduced on two counts.

Surface roughness has the effect of making the contact angle further removed from 90°. If θ is less than 90°, the liquid will penetrate and fill up most of the hollows and pores in the solid and so form a plane surface which is effectively part solid and part liquid; since liquid has zero θ with liquid, θ will therefore decrease. On the other hand, if θ is greater than 90°, the liquid will tend not to penetrate into the hollows and pores in the solid and can, therefore, be regarded as resting on a plane surface which is effectively part solid and part air; since there is practically no adhesion between the liquid and the entrapped air, θ will increase. Surface roughness is a possible cause of contact angle hysteresis.

The method used to prepare the solid surface may affect the contact angle; for example, substances which have crystallised in contact with water often have a lower contact angle than if they crystallise in air, owing to the orientation of the water-attracting groups outwards in the former case. Penetration and entrapment of traces of water in the surface layers, which decreases θ, is also probable in these circumstances.

Wetting agents

Surface-active materials are used as wetting agents in many practical situations. For example, in dips for sheep and cattle and in the application of insecticide and horticultural sprays,[80] the surfaces in question tend to be greasy or wax-like, thus presenting unfavourable conditions for satisfactory surface coverage unless a wetting agent is incorporated. However, in these cases complete wetting is not desirable either since it causes over-efficiency in the drainage of excess liquid from the surface. Wetting agents also find considerable application in the textile industry.

In addition to lowering $\gamma_{L/A}$, it is important that the wetting agent lowers $\gamma_{S/L}$, so that a choice of surfactant to suit the particular nature of the solid surface has to be made (side-effects such as toxicity, foaming, etc., must also be borne in mind). Irregularly shaped surfactant molecules, e.g. sodium di-*n*-octyl sulphosuccinate (Aerosol OT), are often very good wetting agents since micelle formation is not favoured owing to steric considerations, thus permitting relatively high concentrations of unassociated surfactant molecules and hence a greater lowering of $\gamma_{L/A}$ and $\gamma_{S/L}$. Non-ionic surfactants are also good wetting agents.

Water repellency

This is the converse of the previous topic, the aim being to make the contact angle as large as possible. Textile fabrics are made water-repellent by coating the threads with a material having a high contact

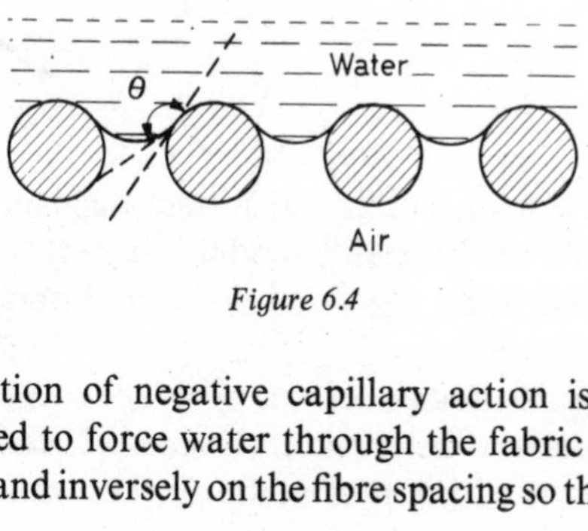

Figure 6.4

angle. A condition of negative capillary action is achieved. The pressure required to force water through the fabric depends on the surface tension and inversely on the fibre spacing so that a moderately tight weave is desirable. The passage of air through the fabric is not hindered.

Examples of water-repelling materials include waxes, petroleum residues, asphalt, soaps of polyvalent metals and silicones. Dimethyldichlorosilane is a very good hydrophobising agent for silica and glass surfaces; it reacts with the —OH groups on the outside of the silicate lattice with the elimination of HCl to give

$$
\begin{array}{c}
\text{CH}_3 \quad \text{CH}_3 \qquad \text{CH}_3 \quad \text{CH}_3 \qquad \text{CH}_3 \quad \text{CH}_3 \\
\diagdown \quad \diagup \qquad\qquad \diagdown \quad \diagup \qquad\qquad \diagdown \quad \diagup \\
\text{Si} \qquad\qquad\qquad \text{Si} \qquad\qquad\qquad \text{Si} \\
\diagup \quad \diagdown \qquad\quad \diagup \quad \diagdown \qquad\quad \diagup \quad \diagdown \\
\text{O} \qquad \text{O} \qquad \text{O} \qquad \text{O} \qquad \text{O} \qquad \text{O} \\
| \qquad\quad | \qquad\quad | \qquad\quad | \qquad\quad | \qquad\quad | \\
-\text{O}-\text{Si}-\text{O}-\text{Si}-\text{O}-\text{Si}-\text{O}-\text{Si}-\text{O}-\text{Si}-\text{O}-\text{Si}-\text{O}- \\
| \qquad\quad | \qquad\quad | \qquad\quad | \qquad\quad | \qquad\quad |
\end{array}
$$

ORE FLOTATION

For a solid particle to float on the surface of a liquid, the total upward pull of the meniscus around it must balance the apparent weight of the particle; for example, a waxed needle can be floated on the surface of water (Figure 6.5) and then sunk by the addition of detergent.

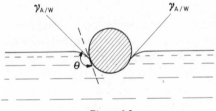

Figure 6.5

The flotation of a solid on a liquid surface depends on the contact angle θ, and since θ can be readily modified by factors such as surface grease, surfactants, etc., the conditions of flotation can also be controlled.[81]

The various constituents of many crude ores have different tendencies to float on the surface of water. These tendencies can be enhanced by the addition of additives, known as collector oils, which adsorb strongly on the metal ore with the result that θ increases to the point where flotation is possible. The collector oils do not adsorb so strongly on siliceous material, which remains wetted by the water and does not float. Organic xanthates and thiophosphates are commonly used as collector oils.

In practice a foaming agent, e.g. crude cresol (soap is unsuitable as it lowers θ too much), is added to the suspension of ground ore and collector oil in water, and air is forced through a fine sieve at the bottom of the vessel. The particles of ore become attached to the air bubbles, which carry them to the surface (Figure 6.6), where they collect as a mineral-rich foam which can be skimmed off.

Contact angles of at least 50° to 75° are required for satisfactory flotation. This can often be achieved with as little as 5 per cent surface coverage, so that the amount of collector oil used is fairly small. Sometimes the ore must be pretreated before it will adsorb the

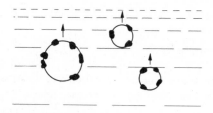

Figure 6.6

additive satisfactorily; for example, zinc sulphide must be pretreated with dilute copper sulphate solution, which deposits copper on the ore surface by electrochemical action. Specificity of flotation may also be achieved by the addition of depressants; for example, cyanide ions prevent ferrous sulphide and zinc sulphide from floating but allow lead sulphide to float. In this way the components of a mixed ore can be separated.

The detailed theory of flotation is a little more complicated than indicated by the above account. Bubble adhesion is maximum when there is only 5–15 per cent monolayer coverage by the collector oil and decreases with further coverage. It is thought that when the bubble and particle interfaces merge, penetration of the film of collector oil around the particle by the film of foaming agent around the bubble occurs. This interlocking between the two films stabilises the air bubble–particle system and is, therefore, most favoured when the particles are only partly covered with a film of collector oil. The function of the foaming agent as such may, therefore, be of secondary importance as the particles themselves act as foam stabilisers (see Chapter 10).

DETERGENCY

Detergency is the theory and practice of dirt removal from solid surfaces by surface chemical means. Soaps have been used as detergents for many centuries. Soap normally consists of the sodium or potassium salts of various long-chain fatty acids and is manufactured by the saponification of glyceride oils and fats (e.g. tallow) with NaOH or KOH, giving glycerol as a by-product

$$
\begin{array}{l}
\underset{|}{CH_2.COOR} \qquad\qquad \underset{|}{CH_2.OH} \qquad R.COONa \\
\underset{|}{CH.COOR'} + 3NaOH = \underset{|}{CH.OH} \ + \ R'.COONa \\
\underset{\text{fat}}{CH_2.COOR''} \qquad\qquad \underset{\text{glycerol}}{CH_2.OH} \qquad \underset{\text{soap}}{R''.COONa}
\end{array}
$$

The potassium soaps tend to be softer and more soluble in water than the corresponding sodium soaps. Soaps from unsaturated fatty acids are softer than those from saturated fatty acids.

Soap is an excellent detergent but suffers from two main drawbacks: (a) it does not function very well in acid solutions because of the formation of insoluble fatty acid, and (b) it forms insoluble precipitates and hence a scum with the Ca^{2+} and Mg^{2+} ions in hard water. Additives such as sodium carbonate, phosphates, etc., help to offset these effects. In the last few decades soap has been partly superseded by the use of synthetic (soapless) detergents, which do not suffer to the same extent from these disadvantages. The alkyl sulphates, alkyl-aryl sulphonates and the non-ionic polyethylene oxide derivatives are perhaps the most important.

Mechanisms of detergency

A satisfactory detergent must possess the following properties.[82]
1. Good wetting characteristics in order that the detergent may come into intimate contact with the surface to be cleaned.

2. Ability to remove or to help remove dirt into the bulk of the liquid.
3. Ability to solubilise or to disperse removed dirt and to prevent it from being redeposited on to the cleaned surface or from forming a scum.

Wetting

The best wetting agents are not necessarily the best detergents, and vice versa. For a homologous series of detergents, such as soaps, alkyl sulphates and alkyl-aryl sulphonates, optimum wetting action is given by about C_8 surfactants, although the longer-chain species are more surface-active. More rapid diffusion to and adsorption at the relevant interfaces by the smaller molecules is probably the reason for this. However, optimum cleaning action is given by about C_{14} surfactants, and for the best all-round performance a chain length of about C_{12} is preferred.[82]

Dirt removal

Dirt is generally of an oily nature with particles of dust, carbon, etc., included. Dirt removal can be considered in terms of the surface

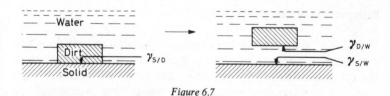

Figure 6.7

energy changes involved. The work of adhesion between a dirt particle and a solid surface (Figure 6.7) is given by

$$W_{S/D} = \gamma_{D/W} + \gamma_{S/W} - \gamma_{S/D} \qquad \ldots \ldots (6.4)$$

The action of the detergent is to lower $\gamma_{D/W}$ and $\gamma_{S/W}$, thus decreasing $W_{S/D}$ and increasing the ease with which the dirt particle can be detached by mechanical agitation.

If the dirt is fluid (oil or grease) its removal can be considered as a contact angle phenomenon. The addition of detergent lowers the contact angle at the triple solid–oil–water interface, as a result of which the oil 'rolls up' (Figure 6.8) and can easily be detached.[83] In this respect, increasing the temperature has a marked effect on detergent efficiency up to about 45°C (most fats melt below this

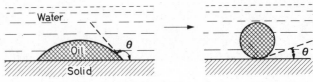

Figure 6.8

temperature) and little effect between about 45°C and just below the boiling point.

It can be seen that surfactants which adsorb at the solid–water and dirt–water interfaces will be the best detergents. Adsorption at the air–water interface with the consequent lowering of surface tension and foaming is, therefore, not necessarily an indication of detergent effectiveness; for example, non-ionic detergents usually have excellent detergent action yet are poor foaming agents and the psychological tendency of the public to correlate these two properties has somewhat restricted their acceptance.

Redeposition of dirt

This can be prevented by the charge and hydration barriers which are set up as a result of detergent molecules being adsorbed on to the cleaned material and on to the dirt particles. Non-ionics are particularly effective in this respect, presumably because of the strong hydration of the polyethylene oxide chains.

The most successful detergents are those forming micelles, and this originally led to the opinion that micelles are directly involved in detergent action, their role probably being that of solubilising oily material. However, detergent action is dependent upon the concentration of unassociated surfactant and practically unaffected by the presence of micelles (other than as a reservoir for replenishing the unassociated surfactant adsorbed from solution).[85] It appears, therefore, that the molecular properties of surfactants associated

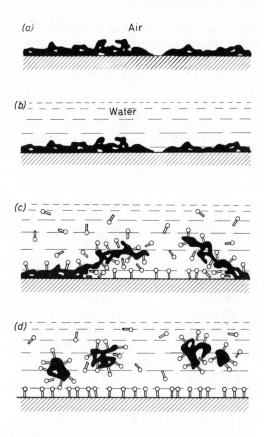

Figure 6.9. Removal of dirt from a solid surface by detergent and mechanical action.[84]

(a) Surface covered with greasy dirt.

(b) Water by itself fails to dislodge the dirt mainly because of its high surface tension and inefficient wetting action.

(c) Detergent is added to the water. The hydrophobic parts of the detergent molecules line up both on the dirt and on the solid surface, thus reducing the adhesion of the dirt to the solid. The dirt may now be dislodged by mechanical action.

(d) Dirt is held suspended in the solution because the detergent molecules form an adsorbed layer on the cleaned surface and around the dirt particles.

(By courtesy of The Scientific American Inc.)

with good detergent action also lead to micelle formation as a competing rather than as a contributing process.[3]

Detergent additives

It is general practice to incorporate 'builders', such as silicates, pyrophosphates and tripolyphosphates, which are not surface-active themselves but which improve the performance of the detergent. Builders fulfil a number of functions, the most important being to sequester (form soluble non-adsorbed complexes) Ca^{2+} and Mg^{2+} ions and act as deflocculating agents, thus helping to avoid scum formation and dirt redeposition. The builders also help by producing the mildly alkaline conditions which are favourable to detergent action.

Sodium carboxymethyl cellulose improves detergent performance in washing textile fabrics, particularly cotton, by forming a protective hydrated adsorbed layer on the cleaned fabric which helps to prevent dirt redeposition. Optical brighteners are commonly incorporated into detergents used for washing textile fabrics. These are fluorescent dyes which absorb ultraviolet light and emit blue light which masks any yellow tint which may develop in white fabrics.

ADSORPTION FROM SOLUTION[86]

To conclude this chapter, some general comments concerning the adsorption of material from solutions on to solid surfaces are appropriate. Adsorption from solution is important in many practical situations, such as those in which modification of the solid surface is of primary concern (e.g. the use of lyophilic material to stabilise dispersions, see page 183) and those which involve the removal of unwanted material from the solution (e.g. the clarification of sugar solutions with activated charcoal). The adsorption of ions from electrolyte solutions and a special case of ion adsorption, that of *ion exchange,* are discussed in Chapter 7. Adsorption processes are, of course, most important in *chromatography*; however, an account of chromatography is not included in this book, (a) because other processes, such as partition and/or molecular sieving, may also be involved to a greater or lesser extent, depending on the type of chromatographic separation being considered, and (b) because chromatography is far too extensive a subject to permit adequate treatment in a relatively small space. Some accounts of chromatographic methods are indicated in References 87 to 91.

Solution adsorption isotherms

Experimentally, the investigation of adsorption from solution is much simpler than that of gas adsorption. A known mass of adsorbent solid is shaken with a known volume of solution at a given temperature until there is no further change in the concentration of the supernatant solution. This concentration can be determined by a variety of methods involving chemical or radiochemical analysis, colorimetry, refractive index, etc. The experimental data are usually expressed in terms of an *apparent adsorption isotherm* in which the amount of solute adsorbed at a given temperature per unit mass of adsorbent [as calculated from the decrease (or increase) of solution concentration] is plotted against the equilibrium concentration.

The theoretical treatment of adsorption from solution, however, is generally more complicated than that of gas adsorption, since adsorption from solution always involves competition between solute(s) and solvent or between the components of a liquid mixture for the adsorption sites. Consider, for example, a binary liquid mixture in contact with a solid. Zero adsorption refers to uniform mixture composition right up to the solid surface, even though (unlike zero gas adsorption) both components are in fact present at the solid surface. If the proportion of one of the components at the surface is greater than its proportion in bulk, then that component is positively adsorbed and, consequently, the other component is negatively adsorbed. Apparent, rather than true, adsorption isotherms are, therefore, calculated from changes in solution concentration. Examples of apparent adsorption isotherms for binary liquid mixtures are given in Figure 6.10. Within the context of certain assumptions, the individual adsorption isotherms can be calculated from the apparent (or composite) adsorption isotherm together with appropriate vapour adsorption data.[8, 9]

Adsorption from solution behaviour can often be predicted qualitatively in terms of the polar/non-polar nature of the solid and

of the solution components. This is illustrated by the isotherms shown in Figure 6.11 for the adsorption of fatty acids from
toluene solution on to silica gel and from aqueous solution on to
carbon.

A polar adsorbent will tend to adsorb polar adsorbates strongly
and non-polar adsorbates weakly, and vice versa. In addition, polar
solutes will tend to be adsorbed strongly from non-polar solvents
(low solubility) and weakly from polar solvents (high solubility),
and vice versa. For the isotherms represented in Figure 6.11b the
solid is polar, the solutes are amphiphilic and the solvent is non-
polar. Fatty acid adsorption is, therefore, strong compared with
that of the solvent. In accord with the above generalisations, the
amount of fatty acid adsorbed at a given concentration decreases
with increasing length of non-polar hydrocarbon chain, i.e. acetic
> propionic > butyric. In Figure 6.11a the solid is non-polar and
the solvent is polar, so, again, fatty acid adsorption is strong
compared with that of the solvent. However, since the adsorbent
is non-polar and the solvent polar, the amount of fatty acid adsorbed at a given concentration now increases with increasing
length of non-polar hydrocarbon chain, i.e. butyric > propionic
> acetic.

Isotherm equations, surface areas

In solution, physical adsorption is far more common than chemisorption. However, chemisorption is sometimes possible; for
example, fatty acids are chemisorbed from benzene solutions on
nickel and platinum catalysts.

Solute adsorption is usually restricted to a monomolecular layer,
since the solid–solute interactions, although strong enough to compete successfully with the solid–solvent interactions in the first
adsorbed monolayer, do not do so in subsequent monolayers. Multilayer adsorption has, however, been observed in a number of cases,
being evident from the shape of the adsorption isotherms and from
the impossibly small areas per adsorbed molecule calculated on the
basis of monomolecular adsorption.

The Langmuir and Freundlich equations (see page 108) are frequently applied to adsorption from solution data, for which they
take the form,

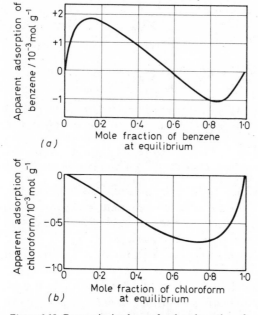

Figure 6.10. Composite isotherms for the adsorption of (*a*) benzene from solution in methanol on to charcoal[92] and (*b*) chloroform from solution in carbon tetrachloride on to charcoal[93] (By courtesy of (*a*) American Chemical Society, (*b*) *Journal of the Chemical Society*)

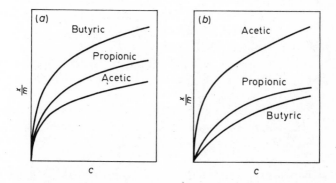

Figure 6.11. Adsorption isotherms for fatty acids; (*a*) from aqueous solutions on to charcoal and (*b*) from toluene solutions on to silica gel

$$\frac{x}{m} = \frac{\left(\dfrac{x}{m}\right)_{max} ac}{1+ac} \qquad \qquad \dots (6.5)$$

and
$$\frac{x}{m} = kc^{1/n} \qquad \qquad \dots (6.6)$$

respectively, where x is the amount of solute adsorbed by a mass m of solid, c is the equilibrium solution concentration and a, k and n are constants.

If the monolayer capacity $\left(\dfrac{x}{m}\right)_{max}$ can be estimated (either directly from the actual isotherm or indirectly by applying the Langmuir equation) and if the effective area occupied by each adsorbed molecule is known, the specific surface area of the solid can be calculated as described on page 115 for gas adsorption.

Since relatively large and asymmetric adsorbate molecules, such as moderately long chain fatty acids and various dyestuffs, are generally involved in adsorption from solution, it is necessary to make assumptions regarding their orientation and packing efficiency in calculating their effective surface coverage. In view of the uncertainties involved in such calculations, it is generally desirable to calibrate a particular adsorption from solution system with the aid of a surface area determined by a less complex method, such as nitrogen adsorption. Adsorption from solution then provides a convenient technique for determining specific surface areas.

7

Charged Interfaces

THE ELECTRIC DOUBLE LAYER

Most substances acquire a surface electric charge when brought into contact with a polar (e.g. aqueous) medium, possible charging mechanisms (elaborated below) being ionisation, ion adsorption and ion dissolution. This surface charge influences the distribution of nearby ions in the polar medium. Ions of opposite charge *(counter-ions)* are attracted towards the surface and (less important) ions of like charge *(co-ions)* are repelled away from the surface. This, together with the mixing tendency of thermal motion leads to the formation of an electric double layer made up of the charged surface and a neutralising excess of counter-ions over co-ions distributed in a diffuse manner in the polar medium. The theory of the electric double layer deals with this distribution of ions and hence with the magnitude of the electric potentials which occur in the locality of the charged surface. This is a necessary first step towards understanding many of the experimental observations concerning the electrokinetic properties, stability, etc., of charged colloidal systems.

Origin of the charge at surfaces

1. *Ionisation*—Proteins acquire their charge mainly through the ionisation of carboxyl and amino groups to give COO^- and NH_3^+

ions. The ionisation of these groups, and so the net molecular charge, depends strongly on the pH of the solution. At low pH a protein molecule will be positively charged and at high pH it will be negatively charged. The pH at which the net charge (and electrophoretic mobility) is zero is called the *iso-electric point* (see Table 2.3 and Figure 7.7).

2. *Ion adsorption*—A net surface charge can be acquired by the unequal adsorption of oppositely charged ions. Ion adsorption may be positive or negative.

Surfaces in contact with aqueous media are more often negatively charged than positively charged. This is a consequence of the fact that cations are generally more hydrated than anions and so have the greater tendency to reside in the bulk aqueous medium, whereas the smaller, less hydrated and more polarising anions have the greater tendency to be specifically adsorbed.

Hydrocarbon oil droplets and even air bubbles suspended in water and in most aqueous electrolyte solutions have negative electrophoretic mobilities (i.e. they migrate towards the anode under the influence of an applied electric field). This net negative charge is explained in terms of negative adsorption of ions. The addition of simple electrolytes, such as NaCl, results in an increase in the surface tension of water (see Figure 4.10) and in the interfacial tension between hydrocarbon oil and water. This is interpreted via the Gibbs equation (page 70) in terms of a negative surface excess ionic concentration. The surface excess concentrations of hydrogen and hydroxyl ions will also be negative. Presumably, cations move away from the air bubble–water and oil–water interfaces more than anions, leaving the kinetic units (which will include some aqueous medium close to the interfaces) with net negative charges.

Preferential negative adsorption of hydrogen ions compared with hydroxyl ions is reflected in the electrophoretic mobility–pH curve for hydrocarbon oil droplets (see Figure 7.7). The magnitude of the electrophoretic mobilities of inert particles such as hydrocarbon oil droplets (*ca.* 0 to -6×10^{-8} m^2 s^{-1} V^{-1}) is comparable with those of simple ions (e.g. $-7 \cdot 8 \times 10^{-8}$ m^2 s^{-1} V^{-1} for Cl$^-$ ions at infinite dilution in aqueous solution at 25°C) which, in view of their relatively large size, reflects a high charge number.

Surfaces which are already charged, e.g. by ionisation, usually show a preferential tendency to adsorb counter-ions, especially of high charge number. It is possible for counter-ion adsorption to cause a reversal of charge.

If surfactant ions are present, their adsorption will usually determine the surface charge.

Hydrated (e.g. protein and polysaccharide) surfaces adsorb ions less readily than hydrophobic (e.g. lipid) surfaces.

3.   *Ion dissolution*—Ionic substances can acquire a surface charge by virtue of unequal dissolution of the oppositely charged ions of which they are composed.

Silver iodide particles in aqueous suspension are in equilibrium with a saturated solution for which the solubility product, $a_{Ag^+} a_{I^-}$, is about 10^{-16} at room temperature. With excess I^- ions the silver iodide particles are negatively charged and with sufficient excess Ag^+ ions they are positively charged. The zero point of charge is not at pAg 8 but is displaced to pAg 5·5 (pI 10·5) because the smaller and more mobile Ag^+ ions are held less strongly than the I^- ions in the silver iodide crystal lattice. The silver and iodide ions are referred to as *potential-determining* ions, since their concentrations determine the electric potential at the particle surface. Silver iodide sols have been used extensively for testing electric double layer and colloid stability theories.

In a similar way, hydrogen and hydroxyl ions are potential-determining for metal oxide and hydroxide sols.

The diffuse double layer

The electric double layer can be regarded generally as consisting of two regions: an inner region which may include adsorbed ions, and a diffuse region in which ions are distributed according to the influence of electrical forces and random thermal motion. The diffuse part of the double layer will be considered first.

Quantitative treatment of the electric double layer presents an extremely difficult and in some respects unsolved problem. The simplest quantitative treatment of the diffuse part of the double layer is that due to Gouy and Chapman, which is based on the following model.

1.   The surface is assumed to be flat, of infinite extent and uniformly charged.
2.   The ions in the diffuse part of the double layer are assumed to

be point charges distributed according to the Boltzmann distribution.

3. The solvent is assumed to influence the double layer only through its dielectric constant, which is assumed to have the same value throughout the diffuse part.

4. A single symmetrical electrolyte of charge number z will be assumed. This assumption facilitates the derivation whilst losing little owing to the relative unimportance of co-ion charge number.

Let the electric potential be ψ_0 at a flat surface and ψ at a distance

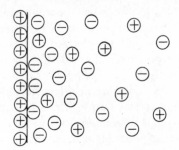

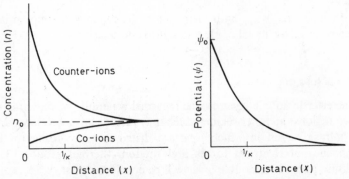

Figure 7.1. Schematic representation of a diffuse electric double layer

x from the surface in the electrolyte solution. Taking the surface to be positively charged (Figure 7.1) and applying the Boltzmann distribution

$$n_+ = n_0 \exp\left[\frac{-ze\psi}{kT}\right]$$

and
$$n_- = n_0 \exp \left[\frac{+ze\psi}{kT} \right]$$

where n_+ and n_- are the respective numbers of positive and negative ions per unit volume at points where the potential is ψ (i.e. where the electric potential energy is $ze\psi$ and $-ze\psi$ respectively), and n_0 is the corresponding bulk concentration of each ionic species.

The net volume charge density ρ at points where the potential is ψ is, therefore, given by

$$\rho = ze \, (n_+ - n_-)$$

$$= zen_0 \left(\exp \left[\frac{-ze\psi}{kT} \right] - \exp \left[\frac{+ze\psi}{kT} \right] \right)$$

$$= -2zen_0 \sinh \frac{ze\psi}{kT} \qquad \qquad \ldots (7.1)$$

ρ is related to ψ by Poisson's equation, which for a flat double layer takes the form

$$\frac{\mathrm{d}^2\psi}{\mathrm{d}x^2} = -\frac{\rho}{\varepsilon} \qquad \qquad \ldots (7.2)$$

where ε is the permittivity.*

Combination of Equations (7.1) and (7.2) gives

$$\frac{\mathrm{d}^2\psi}{\mathrm{d}x^2} = \frac{2zen_0}{\varepsilon} \sinh \frac{ze\psi}{kT} \qquad \qquad \ldots (7.3)$$

The solution[94] of this expression, with the boundary conditions ($\psi = \psi_0$ when $x = 0$; and $\psi = 0$, $\mathrm{d}\psi/\mathrm{d}x = 0$ when $x = \infty$) taken into account, can be written in the form

$$\psi = \frac{2kT}{ze} \ln \left(\frac{1 + \gamma \exp [-\kappa x]}{1 - \gamma \exp [-\kappa x]} \right) \qquad \qquad \ldots (7.4)$$

where
$$\gamma = \frac{\exp [ze\psi_0/2kT] - 1}{\exp [ze\psi_0/2kT] + 1} \qquad \qquad \ldots (7.5)$$

---

* The permittivity of a material is the constant ε in the rationalised expression, $F = \frac{q_1 q_2}{4\pi\varepsilon r^2}$, where F is the force between charges q_1 and q_2 separated by a distance r. The permittivity of a vacuum ε_0 according to this definition is equal to $8{\cdot}854 \times 10^{-12}$ $\mathrm{kg}^{-1} \, \mathrm{m}^{-3} \, \mathrm{s}^4 \, \mathrm{A}^2$. The dielectric constant of a material is equal to the ratio of its permittivity to the permittivity of a vacuum, and is a dimensionless quantity.

and $$\kappa = \left(\frac{2e^2 n_0 z^2}{\varepsilon kT}\right)^{\frac{1}{2}} = \left(\frac{2e^2 N_A cz^2}{\varepsilon kT}\right)^{\frac{1}{2}} \qquad \ldots \ (7.6)$$

where N_A is Avogadro's constant and c is the concentration of electrolyte.

If $ze\psi_0/2kT \ll 1$ $(kT/e = 25\cdot6\,\text{mV}$ at $25°\text{C})$, the Debye-Hückel approximation

$$\left(\exp\left[\frac{ze\psi_0}{2kT}\right] \approx 1 + \frac{ze\psi_0}{2kT}\right)$$

can be made and Equations (7.4) and (7.5) simplify to

$$\psi = \psi_0 \exp\left[-\kappa x\right] \qquad \ldots \ (7.7)$$

showing that at low potentials the potential decreases exponentially with distance from the charged surface. Close to the charged surface, where the potential is likely to be relatively high and the Debye-Hückel approximation inapplicable, the potential is predicted to decrease at a greater than exponential rate.

The potential ψ_0 can be related to the charge density σ_0 at the surface by equating the surface charge with the net space charge in the diffuse part of the double layer $\left(\text{i.e. } \sigma_0 = -\int_0^\infty \rho \, dx\right)$ and applying the Poisson-Boltzmann distribution. The resulting expression is

$$\sigma_0 = (8n_0 \varepsilon kT)^{\frac{1}{2}} \sinh \frac{ze\psi_0}{2kT} \qquad \ldots \ (7.8)$$

which at low potentials reduces to

$$\sigma_0 = \varepsilon \kappa \psi_0 \qquad \ldots \ (7.9)$$

The surface potential ψ_0, therefore, depends on both the surface charge density σ_0 and (through κ) on the ionic composition of the medium. If the double layer is compressed (i.e. κ increased), then either σ_0 must increase, or ψ_0 must decrease, or both.

The potential ψ_0 at the surface of a silver iodide particle depends on the concentration of silver (and iodide) ions in solution. Addition of inert electrolyte increases κ and results in a corresponding increase of surface charge density caused by the adsorption of sufficient potential-determining silver (or iodide) ions to keep ψ_0 approximately constant. In contrast, the charge density at an ionogenic surface remains constant on addition of inert electrolyte (provided that the extent of ionisation is unaffected) and ψ_0 decreases.

From Equation (7.9) it can be seen that, at low potentials, a diffuse double layer has the same capacity as a parallel plate condenser with a distance $1/\kappa$ between the plates. It is customary to refer to $1/\kappa$ (the distance over which the potential decreases by an exponential factor at low potentials) as the 'thickness' of the diffuse double layer.

For an aqueous solution of a symmetrical electrolyte at 25°C, Equation (7.6) becomes

$$\kappa = 0.328 \times 10^{10} \left(\frac{cz^2}{\text{mol dm}^{-3}} \right)^{\frac{1}{2}} \text{m}^{-1} \quad \ldots (7.10)$$

For a 1–1 electrolyte the double layer thickness is, therefore, about 1 nm for a 10^{-1} mol dm^{-3} solution and about 10 nm for a 10^{-3} mol dm^{-3} solution. For unsymmetrical electrolytes the double layer thickness can be calculated by taking z to be the counter-ion charge number.

The inner part of the double layer

The treatment of the diffuse double layer outlined in the last section is based on an assumption of point charges in the electrolyte medium. The finite size of the ions will, however, limit the inner boundary of the diffuse part of the double layer, since the centre of an ion can only approach the surface to within its hydrated radius without becoming specifically adsorbed. Stern proposed a model in which the double layer is divided into two parts separated by a plane (the Stern plane) located at about a hydrated ion radius from the surface, and also considered the possibility of specific ion adsorption.

Specifically adsorbed ions are those which are attached (albeit temporarily) to the surface by electrostatic and/or van der Waals forces strongly enough to overcome thermal agitation. They may be dehydrated, at least in the direction of the surface. The centres of any specifically adsorbed ions are located in the Stern layer, i.e. between the surface and the Stern plane. Ions with centres located beyond the Stern plane form the diffuse part of the double layer, for which the Gouy-Chapman treatment outlined in the previous section, with ψ_0 replaced by ψ_δ, is considered to be applicable.

The potential changes from ψ_0 (the surface or wall potential) to ψ_δ (the Stern potential) in the Stern layer, and decays from ψ_δ to zero in the diffuse double layer.

In the absence of specific ion adsorption the charge densities at

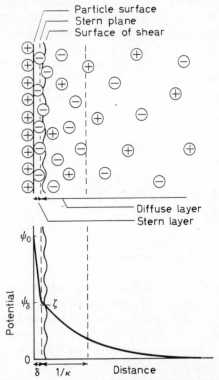

Figure 7.2. Schematic representation of the structure of the electric double layer according to Stern's theory

the surface and at the Stern plane are equal and the capacities of the Stern layer (C_1) and of the diffuse layer (C_2) are given by

$$C_1 = \frac{\sigma_0}{\psi_0 - \psi_\delta} \quad \text{and} \quad C_2 = \frac{\sigma_0}{\psi_\delta}$$

from which

$$\psi_\delta = \frac{C_1 \psi_0}{C_1 + C_2} \qquad \ldots . (7.11)$$

When specific adsorption takes place, counter-ion adsorption generally predominates over co-ion adsorption and a typical double layer situation would be that depicted in Figure 7.2. It is possible, especially with polyvalent or surface active counter-ions, for reversal of charge to take place within the Stern layer, i.e. for ψ_0 and ψ_δ to have opposite signs (Figure 7.3a). Adsorption of surface active

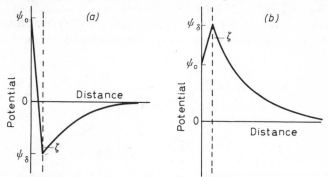

Figure 7.3. (a) Reversal of charge due to the adsorption of surface-active or polyvalent counter-ions. (b) Adsorption of surface-active co-ions

co-ions could create a situation in which ψ_δ has the same sign as ψ_0 and is greater in magnitude (Figure 7.3b).

Stern assumed that a Langmuir-type adsorption isotherm could be used to describe the equilibrium between ions adsorbed in the Stern layer and those in the diffuse part of the double layer. Considering only the adsorption of counter-ions, the surface charge density σ_1 of the Stern layer is given by the expression

$$\sigma_1 = \frac{\sigma_m}{1+\dfrac{N_A}{n_0 V_m} \exp\left[\dfrac{ze\psi_\delta+\phi}{kT}\right]} \qquad \dots (7.12)$$

where σ_m is the surface charge density corresponding to a monolayer of counter-ions, N_A is Avogadro's constant and V_m is the molar volume of the solvent. The adsorption energy is divided between electrical ($ze\psi_\delta$) and van der Waals (ϕ) terms.

Treating the Stern layer as a molecular condenser of thickness δ and with a permittivity ε'

$$\sigma_0 = \frac{\varepsilon'}{\delta} (\psi_0 - \psi_\delta) \qquad \dots (7.13)$$

where σ_0 is the charge density at the particle surface.

For overall electrical neutrality throughout the whole of the double layer

$$\sigma_0 + \sigma_1 + \sigma_2 = 0 \qquad \ldots . (7.14)$$

where σ_2 is the surface charge density of the diffuse part of the double layer and is given by Equation (7.8) with the sign reversed and with ψ_0 replaced by ψ_δ.

Substituting from Equations (7.13), (7.12) and (7.8) into Equation (7.14) gives a complete expression for the Stern model of the double layer:

$$\frac{\varepsilon'}{\delta}(\psi_0 - \psi_\delta) + \frac{\sigma_m}{1 + \dfrac{N_A}{n_0 V_m} \exp\left[\dfrac{ze\psi_\delta + \phi}{kT}\right]}$$
$$- (8n_0 \varepsilon kT)^{\frac{1}{2}} \sinh \frac{ze\psi_\delta}{2kT} = 0 \qquad \ldots . (7.15)$$

This expression contains a number of unknown quantities; however, as indicated below, some information can be derived about these from other sources.

1. *Permittivity of the Stern layer*—The total capacity C of the double layer has been determined from electrocapillary measurements for mercury–aqueous electrolyte interfaces,[95] and from potentiometric titration measurements for silver iodide–aqueous electrolyte interfaces.[96] If the double layer is treated as two capacitors in series then

$$\frac{1}{C} = \frac{1}{C_1} + \frac{1}{C_2}$$

The capacity C_2 of the diffuse part of the double layer can be calculated. At low potentials [see Equation (7.9)].

$$C_2 = \frac{\sigma_2}{\psi_\delta} = \varepsilon\kappa$$
$$= 2\cdot 28 \left(\frac{cz^2}{\text{mol dm}^{-3}}\right)^{\frac{1}{2}} \text{F m}^{-2} \qquad \text{for aqueous electrolyte at 25°C}$$

The capacity of the Stern layer $\left(C_1 = \dfrac{\varepsilon'}{\delta}\right)$ does not depend on electrolyte concentration except insofar that ε' is affected. In the case of the silver iodide–aqueous electrolyte interface, Stern layer capacities of *ca.* $0.1 \, F \, m^{-2}$ to $0.2 \, F \, m^{-2}$ have been calculated; taking $\delta = 5 \times 10^{-10}$ m, this corresponds to a dielectric constant in the Stern layer of *ca.* 5 to 10, which compared with the normal value of *ca.* 80 for water, suggests considerable ordering of water molecules close to the surface.

2. *Stern potentials and electrokinetic (zeta) potentials*—ψ_δ can be estimated from electrokinetic measurements. Electrokinetic behaviour (discussed in the following sections of this chapter) depends on the potential at the surface of shear between the charged surface and the electrolyte solution. This potential is called the *electrokinetic* or ζ *(zeta)* potential. The exact location of the shear plane (which, in reality, is a region of rapidly changing viscosity) is another unknown feature of the electric double layer. In addition to ions in the Stern layer, a certain amount of solvent will probably be bound to the charged surface and form a part of the electrokinetic unit. It is, therefore, reasonable to suppose that the shear plane is usually located at a small distance further out from the surface than the Stern plane and that ζ is, in general, marginally smaller in magnitude than ψ_δ (see Figures 7.2 and 7.3). In tests of double layer theory it is customary to assume identity of ψ_δ and ζ, and the bulk of experimental evidence suggests that errors introduced through this assumption are generally small, especially at lyophobic surfaces. Any difference between ψ_δ and ζ will clearly be most pronounced at high potentials ($\zeta = 0$ when $\psi_\delta = 0$), and at high electrolyte concentration (compression of the diffuse part of the double layer will cause more of the potential drop from ψ_δ to zero to take place within the shear plane). The adsorption of non-ionic surfactant would result in the surface of shear being located at a relatively large distance from the Stern plane and a zeta potential significantly lower that ψ_δ.

3. *Surface potentials*—For an interface such as silver iodide–electrolyte solution the electric potential difference between the solid interior and the bulk solution varies according to the Nernst equation

$$\frac{d\phi}{d(pAg)} = \frac{-2.303 \, RT}{F} (= -59 \text{ mV at } 25°C)$$

ϕ is made up of two terms, ψ_0 and χ. Changes in the χ *(chi)* potential arise from the adsorption and/or orientation of dipolar (e.g. solvent) molecules at the surface or from the displacement of oriented dipolar molecules from the surface. Such effects are difficult to estimate. It is often assumed that χ remains constant during variations of the surface potential and that an expression such as

$$\frac{d\psi_0}{d(pAg)} = \frac{-2\cdot303\,RT}{F} \qquad \dots\,(7.16)$$

is justified. It is also assumed in this expression that no double layer occurs within the solid; this may not be so, since the excess Ag^+ (or I^-) ions of the silver iodide particle do not necessarily all reside at the particle surface. The zero point of zeta potential (which is a measurable quantity, pAg 5·5 at 25°C for AgI in aqueous dis-

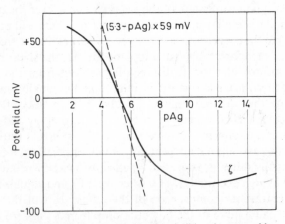

Figure 7.4. Zeta potentials for silver iodide sols prepared by simple mixing.[97] ζ calculated from Smoluchowski equation (By courtesy of Elsevier Publishing Company)

persion medium) can be identified with $\psi_0 = 0$ if specific adsorption of non-potential-determining ions is assumed to be absent. ψ_0 can, therefore, be calculated for a given pAg on the basis of the assumptions outlined in this paragraph.

Experimentally $\left(\dfrac{d\zeta}{d(pAg)}\right)_{\zeta \to 0}$ is found to be about $-40\,mV$ at 25°C for silver iodide[97] (Figure 7.4) and silver bromide[98] hydrosols prepared by simple mixing, and not $-59\,mV$. Assuming that ζ can be identified with ψ_δ and making the assumptions outlined in the last paragraph

$$\left(\frac{d\zeta}{d(pAg)}\right)_{\zeta \to 0} = \left(\frac{d\psi_0}{d(pAg)}\right)\left(\frac{d\psi_\delta}{d\psi_0}\right)_{\zeta \to 0}$$

$$= -59\frac{C_1}{C_1+C_2}\, mV \qquad \ldots (7.17)$$

Values of $\left(\dfrac{d\zeta}{d(pAg)}\right)_{\zeta \to 0}$ calculated from double layer capacities using the above expression seem to be, at least semiquantitatively, in accord with the experimental values (from electrophoretic measurements),[99] suggesting that the assumptions involved may be valid. However, this is a topic which has not as yet been subjected to sufficient systematic investigation for firm conclusions to be drawn.

From the discussion so far it can be appreciated that the Stern model of the electric double layer presents only a rough picture of what is undoubtably a most complex situation. Nevertheless, it provides a good basis for interpreting, at least semiquantitatively, most experimental observations connected with electric double layer phenomena. In particular, it helps to account for the magnitude of electrokinetic potentials (rarely in excess of 75 mV) compared with thermodynamic potentials (which can be several hundred millivolts).

A refinement of the Stern model has been proposed by Grahame[95] who distinguishes between an 'outer Helmholtz plane' to indicate the closest distance of approach of hydrated ions (i.e. the same as the Stern plane), and an 'inner Helmholtz plane' to indicate the centres of ions, particularly anions, which are dehydrated (at least in the direction of the surface) on adsorption.

Finally, both the Gouy-Chapman and the Stern treatments of the double layer assume a uniformly charged surface. The surface charge, however, is not 'smeared out' but is located at discrete sites on the surface. When an ion is adsorbed into the inner Helmholtz plane it will rearrange neighbouring surface charges and, in doing so, impose a self-atmosphere potential ϕ_β on itself (a two-dimensional analogue of the self-atmosphere potential occurring in the Debye-Hückel theory of strong electrolytes). This 'discreteness of of charge' effect can be incorporated into the Stern-Langmuir expression, which now becomes

$$\sigma_1 = \frac{\sigma_m}{1+\dfrac{N_A}{n_0 V_m}\exp\left[\dfrac{ze(\psi_\delta+\phi_\beta)+\phi}{kT}\right]} \qquad \ldots (7.18)$$

The main consequence of including this self-atmosphere term is that the theory now predicts that, under suitable conditions, ψ_δ goes through a maximum as ψ_0 is increased. The discreteness of charge effect, therefore, explains, at least qualitatively, the experimental observations that both zeta potentials (see Figure 7.4) and flocculation concentrations (see Chapter 8) for sols such as silver halides go through a maximum as the surface potential is increased.[100]

Ion exchange

Ion exchange involves an electric double layer situation in which two kinds of counter-ions are present, and can be represented by the equation

$$RA + B = RB + A$$

where R is a charged porous solid. Counter-ions A and B compete for position in the electric double layer around R and, in this respect, concentration and charge number are of primary importance. R may be a cation exchanger (fixed negatively charged groups, such as $-SO_3^-$ or $-COO^-$) or an anion exchanger (fixed positively charged groups, such as $-NH_3^+$). A range of highly porous synthetic cation and anion exchange resins are available commercially. The porosity of the resin facilitates fairly rapid ion exchange.

The most important applications of ion exchange are the softening of water and the 'deionisation' of water.

In the first of these processes, hard water is passed through a column of a cation exchange resin usually saturated with sodium counter-ions. The doubly charged (and, therefore, more strongly adsorbed) calcium ions in the water exchange with the singly charged sodium ions in the resin, thus softening the water. Regeneration of the resin is effected by passing a strong solution of sodium chloride through the column.

The 'deionisation' of water involves both anion and cation exchange. A cation exchange resin saturated with hydrogen ions and an anion exchange resin saturated with hydroxyl ions are used, often in the form of a mixed ion exchange resin. These hydrogen and hydroxyl ions exchange with the cations and anions in the water sample and combine to form water.

Ion exchange has many preparative and analytical uses; for example, the separation of the rare earths is usually achieved by cation exchange followed by elution of their complexes with citric acid.

ELECTROKINETIC PHENOMENA

Electrokinetic is the general description applied to four phenomena which arise when attempts are made to shear off the mobile part of the electric double layer from a charged surface.

If an electric field is applied tangentially along a charged surface a force is exerted on both parts of the electric double layer. The charged surface (plus attached material) tends to move in the appropriate direction, whilst the ions in the mobile part of the double layer show a net migration in the opposite direction carrying solvent along with them, thus causing its flow. Conversely, an electric field is created if the charged surface and the diffuse part of the double layer are made to move relative to one another.

The four electrokinetic phenomena are as follows:

1. *Electrophoresis*—the movement of a charged surface plus attached material (i.e. dissolved or suspended material) relative to stationary liquid by an applied electric field.

2. *Electro-osmosis*—the movement of liquid relative to a stationary charged surface (e.g. a capillary or porous plug) by an applied electric field (i.e. the complement of electrophoresis). The pressure necessary to counterbalance electro-osmotic flow is termed the *electro-osmotic pressure*.

3. *Streaming potential*—the electric field which is created when liquid is made to flow along a stationary charged surface (i.e. the opposite of electro-osmosis).

4. *Sedimentation potential*—the electric field which is created when charged particles move relative to stationary liquid (i.e. the opposite of electrophoresis).

Electrophoresis has the greatest practical applicability of these electrokinetic phenomena and has been studied extensively in its various forms, whereas electro-osmosis and streaming potential have been studied to a moderate extent and sedimentation potential rarely owing to experimental difficulties.

Electrophoresis[101]

A number of techniques have been developed for studying the migration of colloidal material in an electric field.

1.   *Particle (microscope) electrophoresis*—If the material under investigation is in the form of a reasonably stable suspension or emulsion containing microscopically visible particles or droplets, then electrophoretic behaviour can be observed and measured

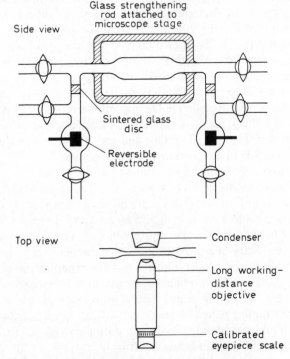

Figure 7.5. A vertically mounted flat particle microelectrophoresis cell[101] (By courtesy of Academic Press Inc.)

directly. Information relevant to soluble material can also be obtained in this way if the substance is adsorbed on to the surface of a carrier, such as oil droplets or silica particles.

The electrophoresis cell usually consists of a horizontal glass tube, of either rectangular or circular cross-section, with an electrode at each end and sometimes with inlet and outlet taps for cleaning and filling (Figures 7.5 and 7.6). Platinum black electrodes are adequate for salt concentrations below about $0.001 \, mol \, dm^{-3}$ to $0.01 \, mol \, dm^{-3}$, otherwise appropriate reversible electrodes, such as $Cu|CuSO_4$ or $Ag|AgCl$, must be used to avoid gas evolution.

Electrophoretic measurements by the microscope method are complicated by the simultaneous occurrence of electro-osmosis.

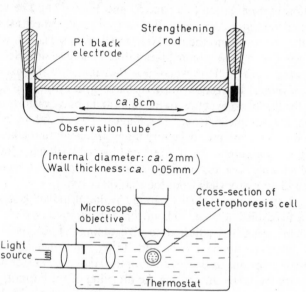

Figure 7.6. Possible arrangement for a thin-walled particle microelectrophoresis cell

The internal glass surfaces of the cell are generally charged, causing an electro-osmotic flow of liquid near to the tube walls together with (since the cell is closed) a compensating return flow of liquid with maximum velocity at the centre of the tube. This results in a parabolic distribution of liquid speeds with depth, and the true electrophoretic velocity is only observed at locations in the tube where the electro-osmotic flow and return flow of the liquid cancel. For a cylindrical cell the 'stationary level' is located at 0.146 of the internal diameter from the cell wall. For a flat cell the 'stationary levels' are located at fractions of about 0.2 and 0.8 of the total depth, the exact locations depending on the width/depth ratio. If the particle and cell surfaces have the same zeta potential the velocity of particles at the centre of the cell is twice their true electrophoretic velocity in a cylindrical cell and 1·5 times their true electrophoretic velocity in a flat cell.

Cylindrical cells are easier to construct and thermostat than flat cells and dark field illumination can be obtained by the ultramicroscopic method of illuminating the sample perpendicular to the direction of observation (see page 53 and Figure 7.6). The volume of dispersion required is generally less for cylindrical cells than for flat cells and, owing to the relatively small cross-section, it is more often possible to use platinum black rather than reversible electrodes with cylindrical cells. However, unless the capillary wall is extremely thin, an optical correction must be made with cylindrical cells to allow for the focusing action of the tube, and optical distortion may prevent measurements from being made at the far stationary level. Cylindrical cells are unsatisfactory if any sedimentation takes place during the measurements; if a rectangular cell is adapted for horizontal viewing (see Figure 7.5), sedimenting particles remain in focus and do not deviate from the stationary levels.

The electrophoretic velocity is found by timing individual particles over a fixed distance (ca. 100 μm) on a calibrated eyepiece scale. The field strength is adjusted to give timings of ca. 10 s—faster times introduce timing errors, and slower times increase the unavoidable error due to Brownian motion. Timings are made at both stationary levels. By alternating the direction of the current, errors due to drift (caused by leakage or convection) can be largely eliminated. The electrophoretic velocity is usually calculated from the average of the reciprocals of about 20 timings.

The potential gradient E at the point of observation is usually calculated from the current I, the cross-sectional area of the channel A and the separately determined conductivity of the dispersion k_0, i.e. $E = I/k_0 A$.

Particle electrophoresis studies have proved to be useful in the investigation of model systems (e.g. silver halide sols and polystyrene latex dispersions) and practical situations (e.g. clay suspensions, water purification and detergency) where colloid stability is involved. In estimating the double layer repulsive forces between particles it is generally assumed that ψ_δ is the operative potential and that ψ_δ and ζ (calculated from electrophoretic mobilities) are identical.

Particle electrophoresis is also a useful technique for character-ising the surfaces of organisms such as bacteria, viruses and blood cells. The nature of the surface charge can be investigated by study-ing the dependence of electrophoretic mobility on factors such as pH, ionic strength, addition of specifically adsorbed polyvalent counter-ions, addition of surface active agents and treatment with

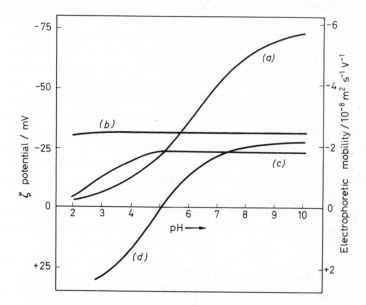

Figure 7.7. Zeta potentials (calculated from electrophoretic mobility data) relating to particles of different ionogenic character plotted as a function of pH in acetate-veronal buffer at a constant ionic strength of $0·05 \text{ mol dm}^{-3}$
(a) Hydrocarbon oil droplets
(b) Sulphonated polystyrene latex particles
(c) Arabic acid (carboxylated polymer) adsorbed on to oil droplets
(d) Serum albumin adsorbed on to oil droplets

specific chemical reagents, particularly enzymes. Figure 7.7 shows, for example, how the mobility–pH curve at constant ionic strength reflects the ionogenic character of some model particle surfaces.

2. *Moving boundary electrophoresis*—An alternative electro-phoretic technique is to study the movement of a boundary formed between a sol or solution and pure dispersion medium. The Tiselius moving boundary method[102] has found wide application, not only

for measuring electrophoretic mobilities, but, particularly, for separating, identifying and estimating dissolved macromolecules, particularly proteins. However, as an analytical technique, where electrophoretic mobilities are not required, moving boundary electrophoresis has been largely superseded by simpler and less expensive zone methods.

The Tiselius cell consists of a U-tube of rectangular cross-section, which is divided into a number of sections built up on ground-glass plates so that they can be moved sideways relative to one another. The protein solution is dialysed against buffer (to avoid subsequent disturbance of the boundary by osmotic flow), and then the sections of the cell are filled with buffered protein solution or buffer solution, for example, as shown in Figure 7.8. Large electrode vessels containing reversible electrodes are then attached, and the whole assembly is immersed in a thermostat. On attaining hydrostatic and thermal equilibrium, the sections of the cell are slid into alignment to form two sharp boundaries. A current is passed through the cell, and the migration of the boundaries is usually followed by a schlieren technique which shows the boundaries as peaks.

The elongated rectangular cross-section of the cell provides a reasonably long optical path for recording the boundary positions,

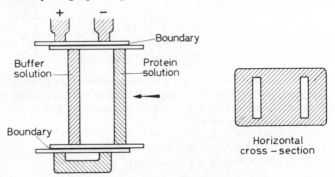

Figure 7.8. A Tiselius electrophoresis cell

and at the same time permits efficient thermostatting. By working at around 0°C to 4°C (where aqueous solutions have maximum density and $d\rho/dT$ is small), convectional disturbance of the boundaries due to the heating effect of the applied current can be minimised even further. The density difference at the boundaries is usually sufficient to prevent disturbance due to electro-osmotic flow at the cell walls.

If the protein solution consists of a number of electrophoretically different fractions, the sharp peak corresponding to the initially formed boundary will broaden and may eventually split up into a number of separate peaks each moving at a characteristic speed. The above precautions against boundary disturbances enable a high degree of resolution to be obtained, thus facilitating the identification, characterisation and estimation of the components in such

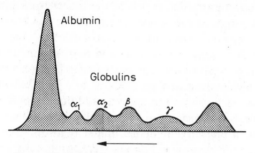

Figure 7.9. Electrophoretic diagram (ascending) for human blood serum

mixtures. For example, the first purpose to which the Tiselius technique was applied[102] was to demonstrate that the component of blood serum once known simply as globulin actually consists of a mixture of several proteins (Figure 7.9).

The moving boundary method is generally complicated by small differences between the ascending and descending boundaries in the two arms of the U-tube. These boundary anomalies result from differences in the conductivity (and, therefore, the potential gradient) at each boundary. They can be minimised by working with low protein concentrations.

3. *Zone electrophoresis*—Zone electrophoresis involves the use of a relatively inert and homogeneous solid or gel framework to support the solution under investigation and minimise convectional disturbances. In addition to being experimentally much simpler than moving boundary electrophoresis, it offers the advantages of giving, in principle, complete separation of all electrophoretically different components and of requiring much smaller samples; however, migration through the stabilising medium is generally a complex process and zone electrophoresis is unsuited for the determination of electrophoretic mobilities.

Zone electrophoresis is used mainly as an analytical technique and, to a lesser extent, for small scale preparative separations. The main applications are in the biochemical and clinical fields, particularly in the study of protein mixtures. Like chromatography, zone electrophoresis is mainly a practical subject and the most important advances have involved improvements in experimental technique and the introduction and development of a range of suitable supporting media. Much of the earlier work involved the use of filter paper as the supporting medium; however, in recent years filter paper has been somewhat superseded by other materials, such as cellulose acetate, starch gel and polyacrylamide gel, which permit sharper separations. The particularly high resolving power of moderately concentrated gel media is to a large extent a consequence of molecular sieving acting as an additional separative factor. For example, blood serum can be separated into about twenty-five components in polyacrylamide gel, but only into five components on filter paper or by moving boundary electrophoresis.

Another recently developed technique, *gel permeation chromatography*,[91] also makes use of this molecular sieving action of gels for fractionating macromolecular material and, by somewhat empirical means, determining relative molecular mass distributions.

Streaming current and streaming potential

The development of a streaming potential when an electrolyte is forced through a capillary or porous plug is, in fact, a complex process, charge and mass transfer occurring simultaneously by a number of mechanisms. The liquid in the capillary or plug carries a net charge (that of the mobile part of the electric double layer) and its flow gives rise to a *streaming current* and, consequently, a potential difference. This potential opposes the mechanical transfer of charge by causing back-conduction by ion diffusion and, to a much lesser extent, by electro-osmosis. The transfer of charge due to these two effects is called the *leak current*, and the measured *streaming potential* relates to an equilibrium condition when streaming current and leak current cancel one another.

Figure 7.10 illustrates a suitable apparatus for studying streaming potentials. To minimise current drain, a very high resistance measuring instrument, such as a vibrating capacitor electrometer, must be used. Most of the difficulties associated with streaming potential measurement originate at the electrodes. A superimposed asymmetry potential often develops; however, by reversing the direction of liquid flow this asymmetry potential can be made to reinforce and oppose the streaming potential and can, therefore, be allowed for.

The streaming current can be measured if the high resistance electrometer is replaced with a microammeter of low resistance

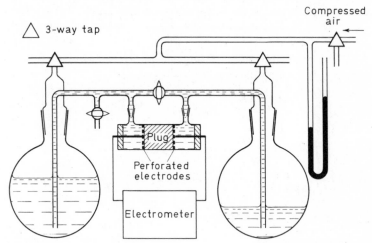

Figure 7.10. A streaming potential apparatus[101] (By courtesy of Academic Press Inc.)

compared with that of the plug. An alternating streaming current can be generated by forcing liquid through the plug by means of a reciprocating pump. The main advantage of studying alternating rather than direct streaming currents is that electrode polarisation is far less likely.

Electro-osmosis

Figure 7.11 illustrates a suitable apparatus for studying electro-osmotic flow through a porous plug. Reversible working electrodes are used to avoid gas evolution. A closed system is employed, the electro-osmotic flow rate being determined by measuring the velocity of an air bubble in a capillary tube (*ca.* 1 mm diameter) which provides a return path for the electrolyte solution.

It may be necessary to correct the experimental data for effects such as electro-osmosis in the measuring capillary tube and electro-osmotic leak-back through the plug.

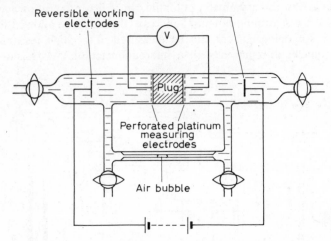

Figure 7.11. An electro-osmosis apparatus[101] (By courtesy of Academic Press Inc.)

ELECTROKINETIC THEORY

Electrokinetic phenomena are only directly related to the nature of the mobile part of the electric double layer and may, therefore, be interpreted only in terms of the zeta potential or the charge density at the surface of shear. No direct information is given about the potentials ψ_0 and ψ_δ (although, as already discussed, the value of ζ may not differ substantially from that of ψ_δ), or about the charge density at the surface of the material in question.

Electrokinetic theory involves both the theory of the electric double layer and that of liquid flow, and is quite complicated. In this section the relation between electrokinetically determined quantities (particularly electrophoretic mobility) and the zeta potential will be considered.

For curved surfaces the shape of the double layer can be described in terms of the dimensionless quantity 'κa', which is the ratio of radius of curvature to double layer thickness. When κa is small, a charged particle may be treated as a point charge; when large, the double layer is effectively flat and may be treated as such.

The Hückel equation (small κa)

Consider κa to be small enough for a spherical particle to be treated as a point charge in an unperturbed electric field, but let the particle be large enough for Stokes' law to apply. Equating the electrical force on the particle with the frictional resistance of the medium

$$Q_E E = 6\pi\eta a v_E$$

or

$$u_E = \frac{v_E}{E} = \frac{Q_E}{6\pi\eta a}$$

where Q_E is the net charge on the particle (i.e. the electrokinetic unit), E the electric field strength, η the viscosity of the medium, a the radius of the particle, v_E the electrophoretic velocity and u_E the electrophoretic mobility.

The zeta potential is the resultant potential at the surface of shear due to the charges $+Q_E$ of the electrokinetic unit and $-Q_E$ of the mobile part of the double layer, i.e.

$$\zeta = \frac{Q_E}{4\pi\varepsilon a} - \frac{Q_E}{4\pi\varepsilon \left(a+\dfrac{1}{\kappa}\right)}$$

$$= \frac{Q_E}{4\pi\varepsilon a(1+\kappa a)} \qquad \ldots \text{(7.19)}$$

where ε is the permittivity of the electrolyte medium (see footnote on page 137). Therefore (neglecting κa compared with unity),

$$u_E = \frac{\zeta\varepsilon}{1\cdot5\,\eta} \qquad \ldots \text{(7.20)}$$

The Hückel equation is not likely to be applicable to particle electrophoresis in aqueous media; for example, particles of radius 10^{-8} m suspended in a 1–1 aqueous electrolyte solution would require an electrolyte concentration as low as 10^{-5} mol dm^{-3} to give $\kappa a = 0\cdot1$. The equation, however, does have possible applicability to electrophoresis in non-aqueous media of low conductance.

The Smoluchowski equation (large κa)

Consider the motion of liquid in the diffuse part of the double layer relative to that of a non-conducting flat surface when an electric field E is applied parallel to the surface. Each layer of liquid will rapidly attain a uniform velocity relative and parallel to the surface with electrical and viscous forces balanced. Equating the electrical and viscous forces on a liquid layer of unit area, thickness dx, distance x from the surface and having a bulk charge density ρ

$$E \rho dx = \left(\eta \frac{dv}{dx}\right)_{x+dx} - \left(\eta \frac{dv}{dx}\right)_x$$

$$= \frac{d}{dx}\left(\eta \frac{dv}{dx}\right) dx$$

Inserting the Poisson equation

$$\left[\rho = -\frac{d}{dx}\left(\varepsilon \frac{d\psi}{dx}\right)\right]$$

$$-E \frac{d}{dx}\left(\varepsilon \frac{d\psi}{dx}\right) = \frac{d}{dx}\left(\eta \frac{dv}{dx}\right)$$

Integrating

$$-E\varepsilon \frac{d\psi}{dx} = \eta \frac{dv}{dx} + \text{Constant}$$

The integration constant is zero, since at $x = \infty$, $d\psi/dx = 0$ and $dv/dx = 0$. Integrating again (assuming that ε and η are constant throughout the mobile part of the double layer)

$$-E\varepsilon\psi = \eta v + \text{Constant}$$

If electrophoresis is being considered, the boundary conditions are $\psi = 0$, $v = 0$ at $x = \infty$ and $\psi = \zeta$, $v = -v_E$ at the surface of shear, where v_E is the electrophoretic velocity, i.e. the velocity of the surface relative to stationary liquid. Therefore,

$$E\varepsilon\zeta = \eta v_E$$

or
$$u_E = \frac{v_E}{E} = \frac{\zeta \varepsilon}{\eta} \qquad \ldots \ldots (7.21)$$

It follows from this expression that the electrophoretic mobility of a non-conducting particle for which κa is large at all points on the surface should be independent of its size and shape provided that the zeta potential is constant.

If electro-osmosis is being considered, a similar expression (i.e. $\frac{v_{E.O.}}{E} = \frac{\zeta\varepsilon}{\eta}$) is derived, the boundary conditions being $\psi = 0$, $v = v_{E.O.}$ at $x = \infty$ and $\psi = \zeta$, $v = 0$ at the surface of shear, where $v_{E.O.}$ is the electro-osmotic velocity.

The Henry equation

Henry[103] derived a general electrophoretic equation for conducting and non-conducting spheres which takes the form

$$u_E = \frac{\zeta\varepsilon}{1 \cdot 5\,\eta}\,[1 + \lambda F(\kappa a)] \qquad \dots (7.22)$$

where $F(\kappa a)$ varies between zero for small values of κa and $1 \cdot 0$ for large values of κa, and $\lambda = (k_0 - k_1)/(2k_0 + k_1)$, where k_0 is the conductivity of the bulk electrolyte solution and k_1 is the conductivity of the particles. For small κa the effect of particle conductance is negligible. For large κa the Henry equation predicts that λ should approach -1 and the electrophoretic mobility approach zero as the particle conductivity increases; however, in most practical cases, 'conducting' particles are rapidly polarised by the applied electric field and behave as non-conductors.

For non-conducting particles ($\lambda = \frac{1}{2}$) the Henry equation can be written in the form

$$u_E = \frac{\zeta\varepsilon}{1 \cdot 5\,\eta}\,f(\kappa a) \qquad \dots (7.23)$$

where $f(\kappa a)$ varies between $1 \cdot 0$ for small κa (Hückel equation) and $1 \cdot 5$ for large κa (Smoluchowski equation) (see Figure 7.12). Zeta potentials calculated from the Hückel equation (for $\kappa a = 0 \cdot 5$) and from the Smoluchowski equation (for $\kappa a = 300$) differ by about 1 per cent from the corresponding zeta potentials calculated from the Henry equation.

The Henry equation is based on several simplifying assumptions:
1. The Debye-Hückel approximation is made.
2. The applied electric field and the field of the electric double layer are assumed to be simply superimposed. Mutual distortion of these fields could affect electrophoretic mobility in two ways (a) through abnormal conductance *(surface*

conductance) in the vicinity of the charged surface, and (b) through loss of double layer symmetry *(relaxation effect)*.

3. ε and η are assumed to be constant throughout the mobile part of the double layer.

Surface conductance

The distribution of ions in the diffuse part of the double layer gives rise to a conductivity in this region which is in excess of that in the bulk electrolyte medium. Surface conductance will affect the distribution of electric field near to the surface of a charged particle and so influence its electrokinetic behaviour. The effect of surface conductance on electrophoretic behaviour can be neglected when κa is small, since the applied electric field is hardly affected by the particle in any case. When κa is not small, calculated zeta potentials may be significantly low on account of surface conductance.

According to Booth[104] and Henry[105] the equation relating electrophoretic mobility with zeta potential for non-conducting spheres with large κa when corrected for surface conductance takes the form

$$u_E = \frac{\zeta \varepsilon}{\eta} \left[\frac{k_0}{k_0 + (k_s/a)} \right] \qquad \dots (7.24)$$

where k_0 is the conductivity of the bulk electrolyte medium and k_s is the surface conductivity.

Substituting ζ_a for $\eta u_E/\varepsilon$ (i.e. ζ_a is the apparent zeta potential calculated from the Smoluchowski equation) gives

$$\frac{1}{\zeta_a} = \frac{1}{\zeta} \left(1 + \frac{k_s}{k_0 a} \right) \qquad \dots (7.25)$$

A plot of $1/\zeta_a$ against $1/a$ should, therefore, give a straight line (if κa is large and if k_s, k_0 and ζ are constant) from which a zeta potential corrected for surface conductance can be obtained by extrapolation. Zeta potentials for oil droplets and protein-covered glass particles have been determined in this way.[106]

The importance of surface conductance at large κa clearly depends on the magnitude of $k_s/(k_0 a)$ compared with unity. The surface conductivity in the mobile part of the double layer can be calculated (and is allowed for in the treatments of relaxation which are outlined in the next section). Experimental surface conductivities (which are not very reliable) tend to be higher than those calculated for the mobile part of the double layer, and the possibility of surface conductance inside the shear plane, especially if the particle surface is porous, has been suggested to account for this discrepancy.[97, 104] There is, therefore, some uncertainty regarding the influence of surface conductance on electrophoretic behaviour; however, it is unlikely to be important when the electrolyte concentration is greater than *ca.* 0·01 mol dm^{-3}.

Relaxation

The ions in the mobile part of the double layer show a net movement in a direction opposite to that of the particle under the influence of

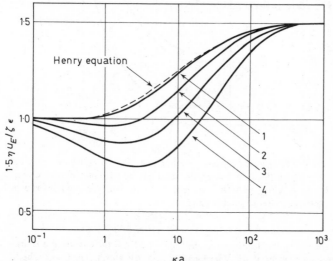

Figure 7.12. Electrophoretic mobility and zeta potential for spherical colloidal particles in 1–1 electrolyte solutions ($\Lambda_+ = \Lambda_- = 70\,\Omega^{-1}\,\text{cm}^2\,\text{mol}^{-1}$). The curves refer to $e\zeta/kT = 1$, 2, 3 and 4 (i.e. $\zeta/\text{mV} = 25\cdot6$, $51\cdot2$, $76\cdot8$ and $102\cdot4$ at 25°C) [After P. H. Wiersema, A. L. Loeb and J. Th. G. Overbeek, *J. Colloid Interface Sci.*, **22**, 78 (1966)]

the applied electric field. This creates a local movement of liquid which opposes the motion of the particle, and is known as *electrophoretic retardation*. It is allowed for in the Henry equation.

The movement of the particle relative to the mobile part of the double layer results in the double layer being distorted, because a finite time (relaxation time) is required for the original symmetry to be restored by diffusion and conduction. The resulting asymmetric mobile part of the double layer exerts an additional retarding force on the particle, known as the *relaxation effect,* and this is not accounted for in the Henry equation. Relaxation can be safely neglected when κa is either small ($<$ *ca.* 0·1) or large ($>$ *ca.* 300),

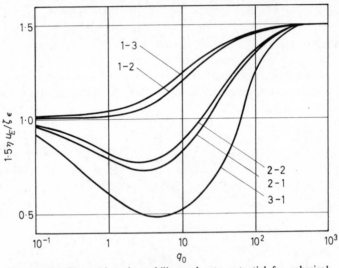

Figure 7.13. Electrophoretic mobility and zeta potential for spherical colloidal particles in electrolyte solutions containing polyvalent ions ($\Lambda_+/z_+ = \Lambda_-/z_- = 70\,\Omega^{-1}\,\mathrm{cm}^2\,\mathrm{mol}^{-1}$). Electrolyte type is numbered with counter-ion charge number first:

$$e\zeta/kT = 2 \text{ in each case}$$
$$q_0 = (2e^2 N_A c z^2/\varepsilon kT)^{\frac{1}{2}} a$$

where c is the electrolyte concentration and z is the counter-ion charge number [After P. H. Wiersema, A. L. Loeb and J. Th. G. Overbeek, *J. Colloid Interface Sci.,* **22**, 78 (1966)]

but it is significant for intermediate values of κa, especially at high potentials and when the counter-ions are of high charge number and/or have low mobilities.

Overbeek[107] and Booth[108] derived equations for spherical particles which allow for retardation, relaxation and surface conductance in the mobile part of the double layer, and which express electrophoretic mobility in terms of a power series in $\dfrac{e\zeta}{kT}$. Owing to mathematical difficulties, these equations were only solved for a restricted number of terms and quantitative validity could only be claimed for $\frac{e\zeta}{kT} < 1$. At higher potentials the relaxation effect was overestimated.

The treatments of Overbeek and Booth have now been superseded both in range of validity and convenience by that of Wiersema, Loeb and Overbeek[109] in which the appropriate differential equations have been solved without approximation using an electronic computer. The main assumptions upon which this treatment is based are

1. The particle is a rigid, non-conducting sphere with its charge uniformly distributed over the surface.
2. The electrophoretic behaviour of the particle is not influenced by other particles in the dispersion.
3. Permittivity and viscosity are constant throughout the mobile part of the double layer, which is described by the classical Gouy-Chapman theory.
4. Only one type each of positive and negative ions are present in the mobile part of the double layer.

Figures 7.12 and 7.13 show the result of a selection of these computations.

Investigations of the electrophoretic behaviour of monodispersed carboxylated polystyrene latex dispersions as a function of particle size and electrolyte concentration by Shaw and Ottewill[110] have confirmed, at least qualitatively, the existence of κa and relaxation effects.

Permittivity and viscosity

Further difficulty in the calculation and interpretation of zeta potentials will arise if the electric field strength $(d\psi/dx)$ close to the shear plane is high enough to significantly decrease ε and/or increase η by dipole orientation. Lyklema and Overbeek[111] examined this problem and concluded that the effect of $d\psi/dx$ on ε is insignificant, but that its effect on η may be significant, especially at high potential and high electrolyte concentration. A significant (and positive) viscoelectric effect would result in the effective location of the shear plane moving farther away from the particle surface with increasing ζ and/or increasing κ, in other words, the physical meaning of the term 'zeta potential' would vary. More recent investigations by Stigter[112] and Hunter[113] suggest, however, that the viscoelectric effect was overestimated by Lyklema and Overbeek and that it is, in fact, insignificant in most practical situations.

Streaming current and streaming potential

The classical equations relating streaming current or streaming potential to zeta potential are derived for the case of a single circular capillary as follows.

Let E_s be the potential difference developed between the ends of a capillary tube of radius a and length l for an applied pressure difference p. Assuming laminar flow, the liquid velocity v_x at a distance x measured from the surface of shear and along a radius of the capillary is given by Poiseuille's equation

$$v_x = \frac{p(2ax - x^2)}{4\eta l}$$

The volume of liquid with velocity v_x can be represented by a hollow cylinder of radius $a - x$ and thickness dx. The rate of flow in this cylindrical layer is, therefore, given by

$$d\left(\frac{dV}{dt}\right) = 2\pi(a - x)v_x dx = \frac{2\pi p(2ax - x^2)(a - x)dx}{4\eta l}$$

The streaming current I_s is given by

$$I_s = \int_0^a \rho\, d\left(\frac{dV}{dt}\right)$$

where ρ is the bulk charge density. If κa is large, the potential decay in the double layer and, therefore, the streaming current are located in a region close to the wall of the capillary tube where x is small compared with a. Substituting for ρ (Poisson's equation, $\frac{d^2\psi}{dx^2} = \frac{-\rho}{\varepsilon}$) and $\frac{dV}{dt}$ (neglecting x compared with a) gives

$$I_s = \frac{\pi\varepsilon p a^2}{\eta l} \int_0^a x \frac{d^2\psi}{dx^2}\, dx$$

The solution of this expression (by partial integration), with the boundary conditions ($\psi = \zeta$ at $x = 0$ and $\psi = 0$, $d\psi/dx = 0$ at $x = a$) taken into account, is

$$I_s = \frac{\varepsilon p A \zeta}{\eta l} \qquad \ldots (7.26)$$

where A is the cross-sectional area of the capillary.

The streaming current and the streaming potential are related by Ohm's law

$$E_s = \frac{I_s l}{k_0 A}$$

where k_0 is the conductivity of the electrolyte solution. Therefore,

$$E_s = \frac{\varepsilon p \zeta}{\eta k_0} \qquad \ldots (7.27)$$

k_0 can be corrected to include a surface conductivity term k_s and Equation (7.27) becomes

$$E_s = \frac{\varepsilon p \zeta}{\eta \left(k_0 + \dfrac{2k_s}{a} \right)} \qquad \ldots (7.28)$$

A more general derivation[97] for porous plugs also leads to Equations (7.26) and (7.27). However, for porous plugs, there is no satisfactory method of correcting streaming potential data for surface conductance. If Equation (7.28) is used with a equal to the average pore radius, the calculated zeta potentials are too low. The importance of surface conductance can be investigated qualitatively by comparing the conductivity ratio for two relevant electrolyte concentrations in bulk and in the plug. A knowledge of surface conductance is not required when relating streaming current to zeta potential. The situation in porous plugs may also be more complicated than accounted for above if (a) the effective area of the plug for the streaming current differs from that for the leak current as a result of the different mechanisms involved, and (b) the plug is compressible and the applied pressure affects the average pore size. The validity of all zeta potentials calculated from streaming (and electro-osmotic) measurements on porous plugs is somewhat dubious.[114]

Electro-osmosis

Experimentally, it is the volume flow rate which is measured. For a single capillary of cross-sectional area A and with large κa, this flow rate is given from the Smoluchowski equation by

$$\frac{dV_{E.O.}}{dt} = Av_{E.O.} = \frac{AE\varepsilon\zeta}{\eta}$$

and since (by Ohm's law) $AE \doteq I/k_0$, where k_0 is the conductivity of the liquid and I the current,

$$\frac{dV_{E.O.}}{dt} = \frac{\varepsilon I\zeta}{\eta k_0} \qquad \ldots\ldots (7.29)$$

or, correcting k_0 to account for the surface conductivity k_s,

$$\frac{dV_{E.O.}}{dt} = \frac{\varepsilon I\zeta}{\eta\left(k_0 + \dfrac{2k_s}{a}\right)} \qquad \ldots\ldots (7.30)$$

Equation (7.29) can be shown by a more general derivation[97] to apply also to porous plugs, but, as in the case of streaming potential, there is no satisfactory method of allowing for surface conductance.

Combination of Equations (7.27) and (7.29) gives a general relationship between electro-osmosis and streaming potential which, again by a more general derivation,[115] can be shown to apply independent of plug structure and surface conductance:

$$\frac{dV_{E.O.}/dt}{I} = \frac{E_s}{p} \qquad \ldots\ldots (7.31)$$

8

Colloid Stability

A most important physical property of colloidal dispersions is the tendency of the particles to aggregate. Encounters between particles dispersed in liquid media occur frequently as a result of Brownian motion and the stability of a dispersion is determined by the interaction between the particles during these encounters.

The principle cause of aggregation is the van der Waals attractive forces between the particles, whereas stability against aggregation is a consequence of repulsive interaction between similarly charged electric double layers and particle-solvent affinity. Particle-solvent affinity promotes stability mainly by mechanical means, which can be considered in terms of the positive desolvation free energy change which accompanies particle aggregation. The adsorption of polymeric material on to the particle surfaces will usually promote stability through increased particle-solvent affinity and by an entropic mechanism, but may induce aggregation by a bridging mechanism.

LYOPHOBIC SOLS

Ideally, lyophobic sols are stabilised entirely by electric double layer interactions; in practice, however, solvation always has some influence on their stability.

Flocculation concentrations—Schulze-Hardy rule

A most notable property of lyophobic sols is their sensitivity to flocculation by small amounts of added electrolyte. The added electrolyte causes a compression of the diffuse parts of the double layers around the particles and may, in addition, exert a specific effect through ion adsorption into the Stern layer. The sol flocculates when the range of double layer repulsive interaction is sufficiently reduced to permit particles to approach close enough for van der Waals forces to predominate.

The flocculation concentration of an indifferent (inert) electrolyte (i.e. the concentration of the electrolyte which is just sufficient to flocculate a lyophobic sol to an arbitrarily defined extent in an arbitrarily chosen time) shows considerable dependence upon the charge number of its counter-ions. In contrast, it is practically independent of the specific character of the various ions, the charge number of the co-ions and the concentration of the sol, and only moderately dependent on the nature of the sol. These generalisations are illustrated in Table 8.1, and are known as the Schulze-Hardy rule.

Table 8.1   FLOCCULATION CONCENTRATIONS IN millimoles PER dm^3 FOR HYDROPHOBIC SOLS[116]

(By courtesy of Elsevier Publishing Company)

As$_2$S$_3$ (−ve sol)		AgI (−ve sol)		Al$_2$O$_3$ (+ve sol)	
LiCl	58	LiNO$_3$	165	NaCl	43·5
NaCl	51	NaNO$_3$	140	KCl	46
KCl	49·5	KNO$_3$	136	KNO$_3$	60
KNO$_3$	50	RbNO$_3$	126		
K acetate	110	(AgNO$_3$	0·01)		
CaCl$_2$	0·65	Ca(NO$_3$)$_2$	2·40	K$_2$SO$_4$	0·30
MgCl$_2$	0·72	Mg(NO$_3$)$_2$	2·60	K$_2$Cr$_2$O$_7$	0·63
MgSO$_4$	0·81	Pb(NO$_3$)$_2$	2·43	K$_2$ oxalate	0·69
AlCl$_3$	0·093	Al(NO$_3$)$_3$	0·067	K$_3$[Fe(CN)$_6$]	0·08
½Al$_2$(SO$_4$)$_3$	0·096	La(NO$_3$)$_3$	0·069		
Al(NO$_3$)$_3$	0·095	Ce(NO$_3$)$_3$	0·069		

The Derjaguin-Landau and Verwey-Overbeek theory

Derjaguin and Landau[117] and Verwey and Overbeek[118] have independently developed a quantitative theory in which the stability of lyophobic sols is treated in terms of the energy changes which take place when particles approach one another. The theory involves estimations of the energy of attraction (London-van der Waals forces) and the energy of repulsion (overlapping of electric double layers) in terms of interparticle distance. Theoretical calculations have been made for the interactions (a) between two parallel charged plates of infinite area and thickness, and (b) between two charged spheres. The calculations for the interaction between flat plates are relevant to the stability of thin soap films, and have been related with a reasonable measure of success to experimental studies in this field[15, 119, 120] (see Chapter 10). The calculations for the inter-action between spheres are relevant to the stability of dispersions and will be outlined. In fact, the conclusions arising from both theoretical treatments are broadly similar.

1. *Double layer repulsive interaction*—The general expression[118, 121] for the repulsive energy V_R which results from the overlapping of the diffuse parts of the double layers around two identical spherical particles (as described by Gouy-Chapman theory) is complex. A relatively simple expression which provides a good approximation for V_R has been given by Reerink and Overbeek,[122] and this will be used in the discussion which follows:

$$V_R = \frac{B\varepsilon k^2 T^2 a\gamma^2}{z^2} \exp\left[-\kappa H\right] \qquad \ldots (8.1)$$

where H is the shortest distance between spheres of radius a, B is a constant equal to $3\cdot93 \times 10^{39}$ A^{-2} s^{-2}, z is the counter-ion charge number and

$$\gamma = \frac{\exp\left[ze\psi_\delta/2kT\right]-1}{\exp\left[ze\psi_\delta/2kT\right]+1} \qquad \ldots (7.5)$$

2. *van der Waals attractive forces*—The forces of attraction between neutral, chemically saturated molecules, postulated by van der Waals, also originate from electrical interactions. Three types of such intermolecular attraction are recognised:

1. Two molecules with permanent dipoles mutually orientate one another in such a way that, on average, attraction results.
2. Dipolar molecules induce dipoles in other molecules so that attraction results.

3. Attractive forces are also operative between non-polar molecules, as is evident from the liquefaction of hydrogen, helium, etc. These universal attractive forces (known as *dispersion forces*) were first explained by London (1930) and are due to the polarisation of one molecule by fluctuations in the charge distribution in a second molecule and vice versa.

With the exception of highly polar materials, London dispersion forces account for nearly all of the van der Waals attraction which is operative. The London attractive energy between two molecules is very short range, varying inversely with the sixth power of the intermolecular distance. For an assembly of molecules, dispersion forces are, to a first approximation, additive and the attractive energy between two particles can be computed by summing the attractions between all interparticle molecule pairs.

The results of such summations predict that the London attractive energy between collections of molecules (e.g. between colloidal particles) decays much less rapidly than that between individual molecules.[123] For the case of two identical spheres of radius a *(in vacuo)* with the shortest distance H between the spheres not greater than *ca.* 10 nm to 20 nm, and when $H \ll a$, the London energy of attraction V_A is given by the approximate expression

$$V_A = \frac{-Aa}{12H} \qquad \qquad \ldots . (8.2)$$

Attraction energies calculated from this equation are likely to be overestimated, especially at large distances ($H > ca.$ 10 nm), owing to a neglect of the finite time required for the propagation of electromagnetic radiation between the particles, the result of which is a weakening of V_A (known as the retardation effect).

Attractive forces between macroscopic bodies have been measured directly by a number of workers; for example, in the first experiment of this kind, Derjaguin and Abricossova[124] used a sensitive electronic feed-back balance to measure the attraction for a plano-convex polished quartz system. The results of these difficult experiments are measured attractive forces which are of the same order of magnitude as those predicted by theory, provided that a retardation correction is made and all residual electric charge is removed from the surfaces in question.

The value of the constant A (known as the Hamaker constant) depends on the nature (particularly the number of atoms per unit volume and the polarisability) of the material of the particles. Its value generally varies between about 10^{-20} J and 10^{-19} J. The presence of liquid dispersion medium, rather than a vacuum (or air), between the particles notably lowers the attraction energy. The constant A in Equation (8.2) must be replaced by an effective Hamaker constant, calculated from the expression

$$A = (\sqrt{A_2} - \sqrt{A_1})^2 \qquad \qquad \dots \text{(8.3)}$$

where subscripts 1 and 2 refer to dispersion medium and particles, respectively. Interparticle attraction will, as might be expected, be

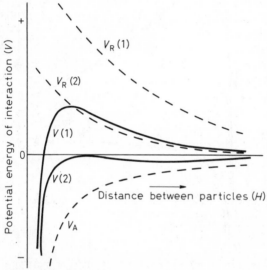

Figure 8.1. Total interaction energy curves ($V(1)$ and $V(2)$), obtained by the summation of an attraction curve, V_A, with different repulsion curves, $V_R(1)$ and $V_R(2)$

weakest when the particles and the dispersion medium are chemically similar, since A_1 and A_2 will be of similar magnitude and the value of A will, therefore, be low. The values of A_1 and/or A_2 are often not known with any great accuracy, especially when solvation may complicate the situation. At the present time, computations of effective Hamaker constants tend to be somewhat approximate (especially when A_1 and A_2 are of similar magnitude). However, notwithstanding the difficulties involved in computing the van der Waals attractive energy between sol particles, most useful theoretical conclusions can still be drawn, even using the approximate Equation (8.2) as a starting point.

3.   *Potential energy curves*—The total energy of interaction is obtained by summation of the attraction and repulsion energies, as illustrated in Figure 8.1. The general character of the resulting

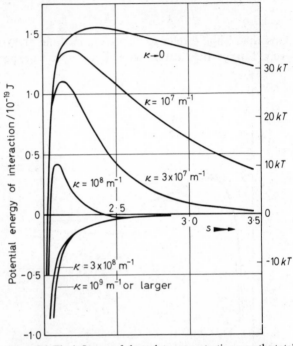

Figure 8.2. The influence of electrolyte concentration κ on the total potential energy of interaction of two spherical particles[118, 121]

$a = 10^{-7}$ m	$T = 298$ K
$A = 10^{-19}$ J	$s = R/a$ (R = distance
$\psi_0 = 25.6$ mV $= \dfrac{kT}{e}$	between centres of spheres)

(By courtesy of Elsevier Publishing Company)

potential energy–distance curve can be deduced from the properties of the two forces. The repulsion energy [Equation (8.1)] is an exponential function of the distance between the particles with a range of the order of the thickness of the electric double layer, and the attraction energy [Equation (8.2)] decreases as an inverse power of the distance between the particles. Consequently, van der Waals attraction will predominate at small* and at large interparticle distances. At intermediate distances double layer repulsion may predominate, depending on the actual values of the two forces. Figure 8.1 shows the two general types of potential energy curve which are possible. The total potential energy curve $V(1)$ shows a repulsive energy maximum, whereas in curve $V(2)$ the double layer repulsion does not predominate over van der Waals attraction at any interparticle distance.

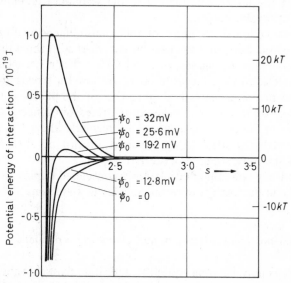

Figure 8.3. The influence of the surface potential ψ_0 on the total potential energy of ineraction of two spherical particles[118, 121]

$a = 10^{-7}$ m $\qquad$ $T = 298$ K
$A = 10^{-19}$ J $\qquad$ $s = R/a$ (R = distance
$\kappa = 10^8$ m^{-1} $\qquad$ between centres of spheres)
(By courtesy of Elsevier Publishing Company)

If the potential energy maximum is large compared with the thermal energy kT of the particles, the system should be stable; otherwise, the system should flocculate. The height of this energy barrier to flocculation depends upon the magnitude of ψ_δ (and ζ) and upon the range of the repulsive forces (i.e. upon $1/\kappa$), as shown in Figures 8.2 and 8.3.

* Repulsion due to overlapping of electron clouds (Born repulsion) predominates at very small distances when the particles come into contact, and so there is a deep minimum in the potential energy curve which is not shown in Figure 8.1.

4.   *Secondary minima*—Another characteristic feature of these potential energy curves is the presence of a secondary minimum at relatively large interparticle distances. If this minimum is moderately deep compared with kT, it should give rise to a loose, easily reversible form of flocculation.* For small particles ($a < ca.$ 10^{-8} m) the secondary minimum is never deep enough for this to happen in those cases where the potential energy maximum is high enough to prevent normal flocculation into the primary minimum. If the particles are larger, flocculation in the secondary minimum may cause observable effects.

Several colloidal systems containing anisodimensional particles, such as iron oxide and tobacco mosaic virus sols, show reversible separation into two phases when the sol is sufficiently concentrated and the electrolyte concentration is too low for irreversible flocculation in the primary minimum. One of the phases is a dilute isotropic sol, the other a more concentrated birefringent sol. The particles in the birefringent phase are regularly aligned as parallel rods or plates $ca.$ 10 nm to 100 nm apart (the distance depending on the pH and ionic strength of the sol).

Secondary minimum flocculation is considered to play an important role in the stability of certain emulsions and foams.

Determination and prediction of flocculation concentrations

The transition between stability and flocculation, although in principle a gradual one, usually occurs over a reasonably small range of electrolyte concentration, and flocculation concentrations can be determined quite sharply. The exact value of the flocculation concentration depends upon the criterion which is set for judging whether the sol is flocculated or not, and this must remain consistent during a series of investigations.

A common method for measuring flocculation concentrations is to prepare a series of about six small test tubes containing equal portions of the sol and to add to each, with stirring, the same volume of electrolyte in concentrations, allowing for dilution, which span the likely flocculation concentration. After standing for a few

---

* Some authors make a distinction between the terms *flocculation* (a secondary minimum effect) and *coagulation* (a primary minimum effect), whereas many others use these terms interchangeably.

minutes an approximate flocculation concentration is noted and a new set of sols is made up with a narrower range of electrolyte concentrations. After standing for a given time (e.g. 2 hours) the sols are reagitated (to break the weaker interparticle bonds and bring small particles into contact with larger ones, thus increasing the sharpness between stability and flocculation), left for a further period (e.g. $\frac{1}{2}$ hour) and then inspected for signs of flocculation. The flocculation concentration can be defined as the minimum electrolyte concentration which is required to produce a visible change in the sol appearance.

An expression for the flocculation concentration of an indifferent electrolyte can be derived by assuming that a potential energy curve such as $V(2)$ in Figure 8.1 can be taken to represent the transition between stability and flocculation into the primary minimum. For such a curve, the conditions $V = 0$ and $\dfrac{dV}{dH} = 0$ hold for the same value of H, i.e. [from Equations (8.1) and (8.2)],

$$V = V_R + V_A = \frac{B\varepsilon k^2 T^2 a\gamma^2}{z^2} \exp\left[-\kappa H\right] - \frac{Aa}{12H} = 0$$

and

$$\frac{dV}{dH} = \frac{dV_R}{dH} + \frac{dV_A}{dH} = -\kappa V_R - \frac{V_A}{H} = 0$$

from which $\kappa H = 1$, therefore,

$$\frac{B\varepsilon k^2 T^2 a\gamma^2}{z^2} \exp\left[-1\right] - \frac{Aa\kappa}{12} = 0$$

giving

$$\kappa_{\text{(flocculation)}} = \frac{4\cdot415\ B\varepsilon k^2 T^2 \gamma^2}{Az^2}$$

Substituting $\left(\dfrac{2e^2 N_A cz^2}{\varepsilon kT}\right)^{\frac{1}{2}}$ for κ [Equation (7.6)] gives

$$c_{\text{(flocculation)}} = \frac{9\cdot75\ B^2 \varepsilon^3 k^5 T^5 \gamma^4}{e^2 N_A A^2 z^6} \qquad \ldots\ldots (8.4)$$

A number of features of the Derjaguin-Landau-Verwey-Overbeek (D.L.V.O.) theory emerge from this expression.

1.   At high potentials γ approaches unity and the theory predicts that the flocculation concentrations of indifferent electrolytes

containing counter-ions with charge numbers 1, 2 and 3 should be in the ratio

$$\frac{1}{1^6} \ : \ \frac{1}{2^6} \ : \ \frac{1}{3^6} \quad \text{or} \quad 100 \ : \ 1{\cdot}6 \ : \ 0{\cdot}13$$

This part of the theory is in strong agreement with experimental evidence (see Table 8.1). Flocculation concentrations would be expected to differ from those otherwise predicted if ψ_δ is sensitive to the flocculating agent, i.e. if specific ion adsorption takes place.

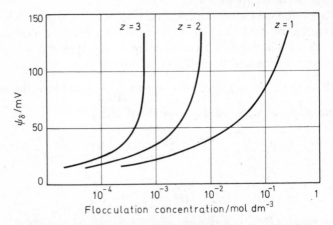

Figure 8.4. Flocculation concentrations calculated from Equation (8.4) taking $A = 10^{-19}$ J, $\varepsilon/\varepsilon_0 = 78{\cdot}5$ and $T = 298$ K, for counter-ion charge numbers 1, 2 and 3. The sol is predicted to be stable above and to the left of each curve and flocculated below and to the right

The addition of potential-determining ions, or counter-ions which are readily adsorbed in the Stern layer, can lead to a reversal of the double layer charge and a second region of stability.

2.   From a typical experimental hydrosol flocculation concentration of $0{\cdot}1$ mol dm^{-3} for $z = 1$ (taking $\gamma = 1$, $T = 298$K and $\varepsilon/\varepsilon_0 = 80$), the effective Hamaker constant A is calculated to be *ca.* 2×10^{-19} J. Although slightly high, this value for A is of the order of magnitude expected from the theory of London-van der Waals forces. Considering the simplified nature of the model which has been used, the agreement between predicted and experimental flocculation concentrations is very good.

3.   Flocculation concentrations are predicted to depend markedly on ψ_δ (and ζ) at low potentials, but to be practically independent of ψ_δ (and ζ) at high potentials (see Figures 8.4 and 8.6).

4.   Flocculation concentrations for spherical particles of a given material at constant ψ_δ should be proportional to ε^3 and independent of the particle size.

The definitions of the term 'flocculation concentration' (a) in relation to experimental measurements, and (b) as a means for arriving at Equation (8.4) are both arbitrary and, no doubt, slightly different from one another. In view of this (in addition to inevitable complications arising from specific ion adsorption and solvation), the results of flocculation concentration measurements can only be taken as support for the validity of the D.L.V.O. theory in its broadest outline. To make more detailed tests of stability theories, study of the kinetics of flocculation presents a better line of approach.

Kinetics of flocculation

The rate at which a sol flocculates depends on the frequency with which the particles encounter one another and the probability that their thermal energy is sufficient to overcome the repulsive potential energy barrier to flocculation when these encounters take place.

The rate at which particles aggregate is given by

$$-\frac{\mathrm{d}n}{\mathrm{d}t} = k_2 n^2$$

where n is the number of particles per unit volume of sol at time t, and k_2 is a second order rate constant.

Integrating, and putting $n = n_0$ at $t = 0$, gives

$$\frac{1}{n} = k_2 t + \frac{1}{n_0} \qquad\qquad \ldots . (8.5)$$

During the course of flocculation k_2 usually decreases and sometimes an equilibrium state is reached with the sol only partially flocculated. This may be a consequence of the height of the repulsion energy barrier increasing with increasing particle size. In experimental tests of stability theories it is usual to restrict measurements to the early stages of flocculation (where the aggregating mechanism is most straightforward) using moderately dilute sols.

The particle concentration during the early stages of flocculation can be determined directly, by visual particle counting, or indirectly, from light scattering measurements.[125, 126] If necessary, flocculation in an aliquot of sol can be halted prior to examination by the addition of a small amount of a stabilising agent, such as gelatin. The rate constant k_2 is given as the slope of a plot of $1/n$ against t.

The potential energy barrier to flocculation can be reduced to zero by the addition of excess electrolyte, thus creating a situation in which every encounter between the particles leads to permanent contact. The theory of rapid (diffusion controlled) flocculation was developed by Smoluchowski.[125] For a monodispersed sol containing spherical particles

$$n = \frac{n_0}{(1+8\pi Dan_0 t)} \qquad \ldots . (8.6)$$

where a is the effective radius of the particles and D is the diffusion coefficient. Substituting $D = kT/6\pi\eta a$ [Equation (2.6)] and combining Equations (8.5) and (8.6) gives

$$k_2^0 = \frac{4kT}{3\eta} \qquad \ldots . (8.7)$$

where k_2^0 is the rate constant for diffusion controlled flocculation.

For a hydrosol at room temperature, the time $t_\frac{1}{2}$ in which the number of particles is halved by diffusion controlled flocculation is calculated from the above equations to be of the order of $10^{11}/n_0$ seconds, if n_0 is expressed in the unit, particles cm^{-3}. If the effect of van der Waals attraction on the collision rate is taken into account, $t_\frac{1}{2}$ is predicted to be lower by a factor of ca. 1·5 to 2. In a typical hydrosol, the number of particles per cm^3 may be about 10^{10} to 10^{11}, and so $t_\frac{1}{2}$ is of the order of a few seconds.

Smoluchowski's theory of rapid flocculation has been found to hold reasonably well for a number of sols when there is sufficient electrolyte present to remove the repulsive potential energy barrier.[125]

When there is a repulsive energy barrier, only a fraction $1/W$ of the encounters between particles lead to permanent contact. W is known as the stability ratio, i.e.

$$W = \frac{k_2^0}{k_2} \qquad \ldots . (8.8)$$

A theoretical expression relating the stability ratio to the potential energy of interaction has been derived by Fuchs[125]

$$W = 2a \int_{2a}^{\infty} \exp\left[\frac{V}{kT}\right] \frac{\mathrm{d}R}{R^2} \qquad \dots \text{(8.9)}$$

Theoretical relationships between the stability ratio and electrolyte concentration can be obtained by numerical solution of this integral

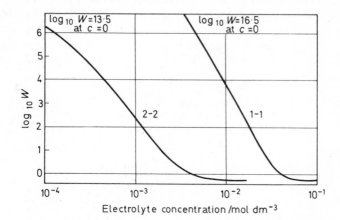

Figure 8.5. Theoretical dependence of stability ratio on electrolyte concentration calculated from Equation (8.9) for $a = 10^{-8}$ m, $A = 2 \times 10^{-19}$ J and $\psi_\delta = 76 \cdot 8$ mV $= 3kT/e$. At high electrolyte concentrations $W < 1$ owing to flocculation being accelerated by van der Waals attractive forces[125] (By courtesy of Elsevier Publishing Company)

for given values of A and ψ_δ. Figure 8.5 shows the results of calculations for 1–1 and 2–2 electrolytes. For constant ψ_δ, a linear relationship between $\log W$ and $\log c$ is predicted for practically the whole of the slow flocculation region.

An alternative approach (which is more convenient, but more approximate) is that of Reerink and Overbeek[122] who have combined an approximate form of Equation (8.9)

$$W \approx \frac{1}{2\kappa a} \exp\left[\frac{V_{\max}}{kT}\right]$$

with Equations (8.1) and (8.2) to derive a theoretical expression which also predicts a linear relationship between $\log W$ and $\log c$ at constant ψ_δ. For a temperature of 25°C and with the particle radius expressed in metres, the resulting equation takes the form

$$\log W = \text{Constant} - 2 \cdot 06 \times 10^9 \left(\frac{a\gamma^2}{z^2}\right) \log c \quad \ldots \ldots (8.10)$$

According to this approximation, $\mathrm{d}\log W/\mathrm{d}\log c$ for the example of $a = 10^{-8}$ m and $\psi_\delta = 3kT/e$ chosen in Figure 8.5 is equal to 9 for 1–1 electrolytes and 4·8 for 2–2 electrolytes, whereas the more

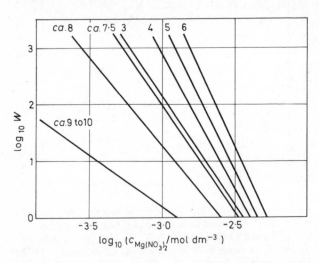

Figure 8.6. Plots of $\log W$ versus $\log c$ for the flocculation of AgI sols at various pI values by magnesium nitrate[127] (By courtesy of Dr. D. Fairhurst and Dr. A. L. Smith)

exact calculations via Equation (8.9) give slopes of 7 and 4·5, respectively.[116]

Flocculation rates have been measured as a function of electrolyte concentration for a number of sols[116, 122, 126, 127] and the predicted linear relationship between $\log W$ and $\log c$ in the slow flocculation region seems to be well confirmed. In addition, the experimental values of $\mathrm{d}\log W/\mathrm{d}\log c$, although somewhat variable, are of the right order of magnitude compared with theoretical slopes.

Figure 8.6 shows some interesting results which have been obtained by Fairhurst and Smith[127] for the flocculation of silver iodide hydrosols at various pI values. As the pI is decreased (and the potential ψ_0 becomes more negative) the slope $d \log W/d \log c$ and the flocculation concentration (which is the concentration which corresponds to an arbitrarily chosen low value of W) increase, as expected, until a pI of about 6 is reached. However, as the pI is reduced below 6, $d \log W/d \log c$ and the flocculation concentration decrease. This apparently anomalous observation [and the corresponding maximum in the zeta potential curve (Figure 7.4)] may be a consequence of the discreteness of charge effect described on page 145.

Experimental data are generally not in accord with the theoretical prediction in Equation (8.10) regarding particle size.[116, 122, 126] For example, Ottewill and Shaw[126] found no systematic variation in $d \log W/d \log c$ for a number of monodispersed carboxylated polystyrene latex dispersions with the particle radius ranging from 30 nm to 200 nm.

Sedimentation volume and gelation

As the particles of a dispersion generally have a density somewhat different from that of the dispersion medium, they will tend to

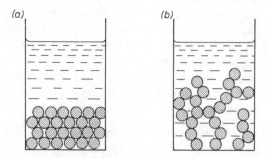

Figure 8.7. Sedimentation volumes for (*a*) deflocculated and (*b*) flocculated particles

accumulate under the influence of gravity at the bottom or at the surface. A sedimentation velocity (see Table 2.2) of up to *ca.* 10^{-8} m s^{-1} is generally counteracted by the mixing tendencies of diffusion and convection. Particle aggregation, of course, enhances sedimentation.

When sedimentation does take place, the volume of the final sediment depends upon the extent of flocculation. Relatively large deflocculated particles pack efficiently to give a dense sediment which is difficult to redisperse, whereas flocculated particles bridge readily and give a loose sediment which is more easily dispersed (see Figure 8.7). In extreme cases, the sedimentation volume may equal the whole volume, and this can lead to the paradoxical situation where a small amount of flocculating agent produces a sediment whilst a larger amount does not. Gentle stirring generally reduces the sedimentation volume.

When the particles flocculate to form a continuous network structure which extends throughout the available volume and immobilises the dispersion medium, the resulting semi-solid system is called a *gel*. The rigidity of a gel depends on the number and the strength of the interparticle links in this continuous structure.

Flocculation and sedimentation volume are important in many practical situations, as illustrated by the following examples.

1. *Agricultural soil*—It is necessary to maintain agricultural soil in a reasonably flocculated state in order to achieve good aeration and drainage, and treatment with flocculants, such as calcium salts (lime or gypsum) or organic polyelectrolytes (so-called soil conditioners) is common practice. An extreme example of the effect of soil deflocculation occurs when agricultural land is flooded with sea water. The calcium ions of the naturally occurring clay minerals in the soil exchange with the sodium ions of the sea water. Subsequent leaching of the sodium ions from the soil by rain water leads to deflocculation and the soil packs into a hard mass which is unsuitable for plant growth. Conversely, water seepage from reservoirs can be reduced by initial flooding with sea water.

2. *Oil well drilling*—In the drilling of oil wells a drilling mud (usually a bentonite clay suspension) is used (a) as a coolant, (b) for removing the cuttings from the bore-hole, and (c) to seal the sides of the bore-hole with an impermeable filter cake. The pumping and sealing features of this operation are most effective if the drilling mud is deflocculated; however, a certain amount of mud rigidity is required to reduce sedimentation of the cuttings, especially during an interruption of circulation. These opposite requirements are somewhat reconciled by maintaining the drilling mud in a partially flocculated, thixotropic (page 198) state. If the drilling mud stiffens,

partial deflocculation can be effected by the addition of a small amount of a peptising agent, such as a polyphosphate. The plate-like particles of clays such as bentonite often have negatively charged faces and positively charged edges when in contact with aqueous media and flocculate quite readily by an edge-to-face mechanism to form a gel structure, even at moderately low clay concentrations.[128] The main function of the polyphosphate is to reverse the positive charge on the edges of the clay particles. The relatively small edge area makes this process economically attractive.

3. *Paints*—The particles in pigmented paints are often large enough to settle even when deflocculated, therefore, it is desirable that they should be flocculated to a certain extent to facilitate re-dispersion.

SYSTEMS CONTAINING LYOPHILIC MATERIAL

In addition to van der Waals attraction and the possibility of electric double layer repulsion, the stability of dispersions which contain surface active material (especially if polymeric) may be considerably influenced by additional factors, such as those involving desorption energy, entropic and bridging effects. These effects are not readily amenable to mathematical treatment or, even, entirely understood, and the present theories relating to the stability of colloidal systems which contain lyophilic material tend to be rather qualitative.

Solvation

Macromolecular solutions are stabilised by a combination of electrical double layer interaction and solvation, and both of these stabilising influences must be sufficiently weakened before precipitation will take place. For example, gelatin has a sufficiently strong affinity for water to be soluble (unless the electrolyte concentration is very high) even at its isoelectric pH, where there is no double layer interaction. Casein, on the other hand, exhibits weaker hydrophilic behaviour and is precipitated from aqueous solution when the pH is near to the isoelectric point.

Owing to their affinity for water, hydrophilic colloids are un-affected by the small amounts of added electrolyte which cause hydrophobic sols to flocculate, but are often precipitated (salted out) when the electrolyte concentration is high. The ions of the added electrolyte dehydrate the hydrophilic colloid by competing for its water of hydration. The salting-out efficacy of an electrolyte, therefore, depends upon the tendencies of its ions to become hydrated. Thus, cations and anions can be arranged in the following lyotropic series of approximately decreasing salting-out power

$$Mg^{2+} > Ca^{2+} > Sr^{2+} > Ba^{2+}$$

$$> Li^+ > Na^+ > K^+ > NH_4^+ > Rb^+ > Cs^+$$

and $$citrate^{3-} > SO_4^{2-} > Cl^- > NO_3^- > I^- > CNS^-$$

Ammonium sulphate, which has a high solubility, is often used to precipitate proteins from aqueous solution.

Lyophilic colloids can also be desolvated (and precipitated if the electrical double layer interaction is sufficiently small) by the addition of non-electrolytes, such as acetone or alcohol to aqueous gelatin solution and petrol ether to a solution of rubber in benzene.

Protective agents

The stability of lyophobic sols can often be enhanced by the addition of soluble lyophilic material which adsorbs on to the particle surfaces. Such adsorbed material is called a *protective agent*. The mechanism of protection is generally complex and a number of factors may be involved.

1.   If the protective agent contains ionisable groups, electric double layer repulsion between the particles may be enhanced.

2.   Adsorbed layers of protective agent may cause a significant lowering of the effective Hamaker constant and, therefore, a weakening of the interparticle van der Waals attraction.

3.   The presence of adsorbed films around the particles may necessitate the expending of an appreciable positive free energy of desorption before the particles can get close enough to one another for van der Waals attraction to predominate.

4.   If macromolecular material is adsorbed on to the particle surfaces such that polymeric chains extend a reasonable distance

into the dispersion medium, interaction between these chains as the particles approach one another is accompanied by a decrease in entropy (i.e. decrease in disorder). Since the enthalpy change is negligible, this interaction involves a positive free energy change ($\Delta G = \Delta H - T\Delta S$) and so opposes particle aggregation. This effect is referred to as *steric* or *entropic stabilisation*.

Protective action is particularly important in non-aqueous dispersions (paints, printing ink, etc.) where electric double layer interaction is likely to be weak or absent. Macromolecular materials are usually the most effective protective agents; in general, they have a relatively high negative free energy of adsorption and can form thick adsorbed layers, thus keeping the particles widely separated. However, it is desirable that macromolecular protective agents should not adsorb too strongly and become attached to the particle surface at too many points along the polymer chains, since this would lead to a relatively thin and less effective stabilising film.

Sensitisation

In certain cases, colloidal dispersions are made more sensitive to precipitation by the addition of small quantities of materials which, if used in larger amounts, would act as protective agents. Several factors may contribute to such observations.

1. If the sol particles and the additive are oppositely charged, sensitisation results when the concentration (and adsorption) of the additive is such that the charge on the particles is neutralised, whereas protection results at higher concentrations because of a reversal of the charge and increasing solvation.

2. At low concentrations, surface-active additives may form a first adsorbed layer on the sol particles with the lyophobic part orientated outwards, thus sensitising the sol. At higher concentrations a second, oppositely orientated, layer would then give protection.[129]

3. Long-chain additives, such as gelatin, can sometimes bring about a rather loose flocculation by a bridging mechanism in which the molecules are adsorbed with part of their length on two or more particles.[130, 131] Such flocculation normally occurs over a narrow

range of additive concentrations; at higher concentrations protective action is obtained, since bridging can occur only through particle collisions under conditions where further adsorption of the additive is possible.

9
Rheology

INTRODUCTION

Rheology is the science of the deformation and flow of matter, and its study has contributed much towards clarifying ideas concerning the nature of colloidal systems. Rheology is also a subject of tremendous and increasing technological importance—in many industries, such as rubber, plastics, food, paint and textiles, the suitability of the products involved is to a large extent judged in terms of their mechanical properties.

The most straightforward rheological behaviour is exhibited on the one hand by Newtonian viscous fluids and on the other by Hookean elastic solids. However, most materials, particularly those of a colloidal nature, exhibit mechanical behaviour which is intermediate between these two extremes with both viscous and elastic characteristics in evidence. Such materials are termed *viscoelastic*.

There are two general approaches to rheology, the first being to set up mathematical expressions which describe rheological phenomena without undue reference to their causes, and the second, with which the following discussion is mainly concerned, is to correlate observed mechanical behaviour with the detailed structure of the material in question. This is not an easy task. The rheological behaviour of colloidal systems is generally very complicated and reflects not only the characteristics of the individual particles but also particle–particle and particle–solvent interactions. The individual molecules or particles may be joined by primary valency bonds of cross-linkage, and/or associate owing to van der Waals forces of attraction, and/or simply become mechanically entangled. Because of the complications involved, this aspect of rheology is still in many respects a mainly descriptive science. However, in recent years considerable advances have been made towards understanding rheological behaviour and putting it on to a quantitative basis.[132]

For convenience, this chapter has been divided into three sections in which the viscosity of dilute solutions and dispersions, non-Newtonian flow, and the viscoelastic properties of semi-solid systems are discussed.

VISCOSITY

Newtonian viscosity

The viscosity of a liquid is a measure of the internal resistance offered to the relative motion of different parts of the liquid. Viscosity is described as Newtonian when the shearing force per unit area τ between two parallel planes of liquid in relative motion is proportional to the velocity gradient dv/dx between the planes, i.e.

$$\tau = \eta dv/dx \qquad \qquad \ldots (9.1)$$

where η is the *coefficient of viscosity*. The dimension of η is, therefore, (mass) (length)$^{-1}$ (time)$^{-1}$.

For most pure liquids and for many solutions and dispersions η is a well defined quantity for a given temperature and pressure which is independent of τ and dv/dx, provided that the flow is streamlined (i.e. laminar). For many other solutions and dispersions, especially if concentrated and/or if the particles are asymmetric, deviations from Newtonian flow are observed. The main causes of non-Newtonian flow are the formation of a structure throughout the system and orientation of asymmetric particles caused by the velocity gradient.

Measurement of viscosity

1. *Capillary flow methods*—The most frequently employed methods for measuring viscosities is based on flow through a capillary tube. The pressure under which the liquid flows furnishes the shearing stress.

The relative viscosities of two liquids can be determined by using a simple Ostwald viscometer (Figure 9.1). Sufficient liquid is introduced into the viscometer for the levels to be at B and C. Liquid

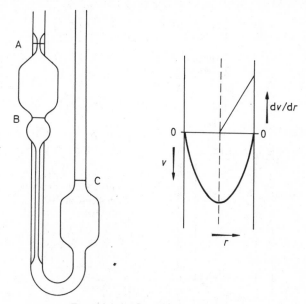

Figure 9.1. An Ostwald viscometer

is then drawn up into the left-hand limb until the liquid levels are above A and at the bottom of the right-hand bulb. The liquid is then released and the time for the left-hand meniscus to pass between the marks A and B is measured.

Since the pressure at any instant driving the liquid through the capillary is proportional to its density

$$\eta = k\rho t \qquad \qquad \dots (9.2)$$

where k is the viscometer constant, ρ the density of the liquid, and t the flow time. Therefore, for two different liquids

$$\frac{\eta_1}{\eta_2} = \frac{\rho_1 t_1}{\rho_2 t_2} \qquad\qquad \dots (9.3)$$

Accurate thermostatting is necessary owing to the marked dependence of viscosities on temperature. Dust and fibrous materials, which might block the capillary, must be removed from the liquid prior to its introduction into the viscometer. A viscometer is selected which gives flow times in excess of *ca.* 100 s, otherwise a kinetic energy correction is necessary.

The capillary method is simple to operate and precise (*ca.* 0·01

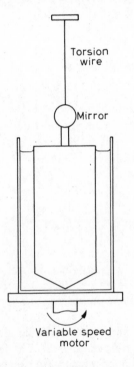

Torsion wire

Mirror

Figure 9.2. A concentric cylinder (Couette) viscometer

Variable speed motor

to 0·1 per cent) in its results, but suffers from the disadvantage that the rate of shear varies from zero at the centre of the capillary to a maximum (which decreases throughout the determination) at the wall. Thus, with asymmetric particles a viscosity determination in an Ostwald viscometer could cover various states of orientation and the measured viscosity, although reproducible, would have little theoretical significance.

2.   *Rotational methods*—Concentric cylinder and cone and plate instruments are particularly useful for studying the flow behaviour of non-Newtonian liquids.

In the first of these techniques an approximation to uniform rate of shear throughout the sample is achieved by shearing a thin film of the liquid between concentric cylinders. The outer cylinder can be rotated (or oscillated) at a constant rate and the shear stress measured in terms of the deflection of the inner cylinder, which is suspended by a torsion wire (Figure 9.2); or the inner cylinder can be rotated (or oscillated) with the outer cylinder stationary and the resistance offered to the motor measured.

At a distance r from the axis of the cylinders (Figure 9.2)

$$\frac{\mathrm{d}v}{\mathrm{d}r} = 2\omega \frac{1/r^2}{1/R_1^2 - 1/R_2^2}$$

where ω is the angular velocity of the outer rotating cylinder, and R_1 and R_2 are the radii of the inner and outer cylinders respectively.

When R_1 and R_2 are not too different the rate of shear across the gap is not too variable and can be controlled by the speed of rotation. The above expression then approximates to

$$\frac{\mathrm{d}v}{\mathrm{d}r} = \frac{\omega R}{d}$$

where d is the gap between the cylinders, and R the mean radius of the cylinders.

The viscous drag on the inner cylinder is $k\theta R$, where k is the torsional constant of the wire and θ is the angular deflection of the cylinder. This force is exerted over an area of $2\pi Rh$, where h is the effective height of liquid in contact with the cylinders. Therefore

$$\eta = \frac{k\theta d}{2\pi h\omega R}$$

or $$\eta = \frac{K\theta}{\omega h} \qquad \qquad \dots\,(9.4)$$

where K is an apparatus constant (usually obtained by calibration with a liquid of known viscosity).

In practice, an end-correction is necessary. If the inner cylinder is appropriately cone-shaped at its end, the liquid at the bottom of the viscometer is sheared at approximately the same rate as the liquid between the cylinder walls, and the end-correction can be included by calibrating the instrument for a given level of liquid. Alternatively, the end-correction can be eliminated by making two measurements with the viscometer filled to different levels, or, for moderately viscous liquids, by having mercury at the bottom of the viscometer and in contact with the lower end of the inner cylinder.

Cone and plate instruments (see Figure 9.3) permit the velocity gradient to be kept constant throughout the sample, and are particularly useful for studying highly viscous materials. A very versatile cone and plate rheometer, known as a rheogoniometer, has been

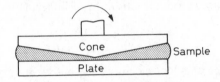

Figure 9.3. Cone and plate method

developed by Weissenberg, which enables both tangential forces and normal forces (i.e. forces which tend to push the cone upwards, see page 204) to be measured either in rotation or in oscillation.

Viscosities of dilute colloidal solutions and dispersions

1. *Functions of viscosity*—When colloidal particles are dispersed in a liquid the flow of the liquid is disturbed and the viscosity is higher than that of the pure liquid. The problem of relating the viscosities of colloidal dispersions (especially when dilute) with the nature of the dispersed particles has been the subject of much experimental investigation and theoretical consideration. In this respect, viscosity increments are of greater significance than absolute viscosities, and the following functions of viscosity are defined:

η_0 = viscosity of pure solvent or dispersion medium

η = viscosity of solution or dispersion

$\eta_{rel} = \eta/\eta_0$ = relative viscosity (or viscosity ratio)

$\eta_{sp} = \eta_{rel} - 1$ = specific increase in viscosity (or viscosity ratio increment)

η_{sp}/c = reduced viscosity (or viscosity number)

$$[\eta] = \lim_{c \to 0} \eta_{sp}/c = \lim_{c \to 0} \frac{1}{c} \ln \eta_{rel}$$

= intrinsic viscosity (or limiting viscosity number)

From the above expressions it can be seen that reduced and intrinsic viscosities have the unit of reciprocal concentration. When considering particle shape and solvation, however, concentration is generally expressed in terms of the volume fraction ϕ of the particles (i.e. volume of particles/total volume) and the corresponding reduced and intrinsic viscosities are, therefore, dimensionless.

2.   *Spherical particles*—Einstein (1906)[21] made a hydrodynamic calculation (under assumptions similar to those of Stokes, see page 17) relating to the disturbance of the flow lines when identical, non-interacting, rigid, spherical particles are dispersed in a liquid medium, and arrived at the expression

$$\eta = \eta_0 (1 + k\phi)$$

where k is a constant equal to 2·5, i.e.

$$\eta_{sp} = 2 \cdot 5 \, \phi \quad \text{or} \quad [\eta]_\phi = 2 \cdot 5 \qquad \ldots (9.5)$$

The effect of such particles on the viscosity of a dispersion depends, therefore, only on the total volume which they occupy and is independent of their size.

The validity of Einstein's equation has been confirmed experimentally for dilute suspensions ($\phi < ca.$ 0·02) of glass spheres, certain spores and fungi, polystyrene particles, etc., in the presence of sufficient electrolyte to eliminate charge effects.[133]

For dispersions of non-rigid spheres (e.g. emulsions) the flow lines may be partially transmitted through the suspended particles, making k in Einstein's equation less than 2·5.

The non-applicability of the Einstein equation at moderate concentrations is mainly due to an overlapping of the disturbed regions of flow around the particles. A number of equations, mostly of the type $\eta = \eta_0 (1 + a\phi + b\phi^2 + \ldots)$, have been proposed to allow for this,[133, 134] the best known being that of Guth and Simha[135]

$$\eta = \eta_0 (1 + 2 \cdot 5 \, \phi + 14 \cdot 1 \, \phi^2 + \ldots)$$

which was derived theoretically. In practice, values of b generally range between about 5 and 8.

3. *Solvation and asymmetry*—The volume fraction term ϕ in viscosity equations must include any solvent which acts kinetically as a part of the particles. The intrinsic viscosity is, therefore, proportional to the solvation factor (i.e. the ratio of solvated and unsolvated volumes of dispersed phase). The solvation factor will generally increase with decreasing particle size.

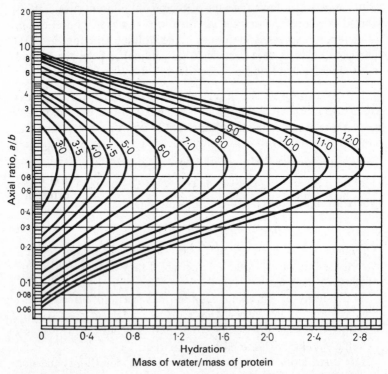

Figure 9.4. Values of axial ratio and hydration compatible with various intrinsic viscosities (contour lines denote $[\eta]_\phi$ values) (By courtesy of the authors[20] and Reinhold Publishing Corporation)

Particle asymmetry has a marked effect on viscosity and a number of complex expressions relating intrinsic viscosity (usually extrapolated to zero velocity gradient to eliminate the effect of orientation) to axial ratio for rods, ellipsoids, flexible chains, etc., have been proposed.[136] For randomly orientated, rigid, elongated particles the intrinsic viscosity is approximately proportional to the square of the axial ratio.

Both asymmetry and solvation increase the intrinsic viscosity. With its application to proteins in mind, Oncley[20] has computed intrinsic viscosities for ellipsoids of revolution of varying degrees of asymmetry and hydration (Figure 9.4—cf. Figure 2.1).

4. *Electroviscous effects*[137]— When dispersions containing charged particles are sheared, extra energy is required to overcome the interaction between ions in the double layers around the particles and the electrical charge on the particle surfaces, thus leading to an increased viscosity.

For charged flexible chains, in addition to the above effect (which is usually small), the nature of the double layer influences the chain configuration. At low ionic strengths the double layer repulsions between the various parts of the flexible chain have a relatively long range and tend to give the chain an extended configuration, whereas at high ionic strengths the range of the double layer interactions is less, thus permitting a more coiled configuration. Therefore, the viscosity decreases with increasing ionic strength, often in a marked fashion.

5. *Polymer relative molecular masses from viscosity measurements*—Viscosity measurements cannot be used to distinguish between particles of different size but of the same shape and degree of solvation. However, if the shape and/or solvation factor alters with particle size, viscosity measurements can be used for determining particle sizes.

If a polymer molecule in solution behaves as a random coil, its average end-to-end distance is proportional to the square root of its extended chain length (see page 20), i.e. proportional to $M_r^{0.5}$, where M_r is the relative molecular mass. The average solvated volume of the polymer molecule is, therefore, proportional to $M_r^{1.5}$ and, since the unsolvated volume is proportional to M_r, the average solvation factor is proportional to $M_r^{1.5}/M_r$ (i.e. $M_r^{0.5}$). The intrinsic viscosity of a polymer solution is, in turn, proportional to the average solvation factor of the polymer coils, i.e.

$$[\eta] = KM_r^{0.5}$$

where K is a proportionality constant.

For most linear high polymers in solution the chains are somewhat more extended than random, and the relation between intrinsic viscosity and relative molecular mass can be expressed by the general equation proposed by Mark and Houwink

$$[\eta] = KM_r^{\alpha} \qquad \ldots (9.6)$$

K and α are constants characteristic of the polymer–solvent system (α depends on the configuration of the polymer chains) and approximately independent of relative molecular mass.[138]

Table 9.1 VALUES OF K AND α FOR SOME POLYMER-SOLVENT SYSTEMS

System	$K/m^3\ kg^{-1}$	α
Cellulose acetate in acetone (25°C)	1.49×10^{-5}	0.82
Polystyrene in toluene (25°C)	3.70×10^{-5}	0.62
Polymethylmethacrylate in benzene (25°C)	0.94×10^{-5}	0.76
Polyvinyl chloride in cyclohexanone (25°C)	0.11×10^{-5}	1.0

In view of experimental simplicity and accuracy, viscosity measurements are extremely useful for routine relative molecular mass determinations on a particular polymer–solvent system. K and α for the system are determined by measuring the intrinsic viscosities of polymer fractions for which the relative molecular masses have been determined independently, e.g. by osmotic pressure, sedimentation, or light scattering.

For polydispersed systems an average relative molecular mass intermediate between number average ($\alpha = 0$) and mass average ($\alpha = 1$) usually results.

NON-NEWTONIAN FLOW

Steady-state phenomena

1. *Shear-thinning*—Shear-thinning, as the term suggests, is characterised by a gradual (time-independent) decrease in apparent viscosity with increasing rate of shear, and can arise from a number of causes.

If particle aggregation occurs in a colloidal system, then an increase in the shear rate will tend to break down the aggregates, which will result, among other things, in a reduction of the amount of solvent immobilised by the particles, thus lowering the apparent viscosity of the system.

Shear-thinning is particularly common to systems containing asymmetric particles. Asymmetric particles disturb the flow lines to a greater extent when they are randomly orientated at low velocity gradients than when they have been aligned at high velocity gradients. In addition, particle interaction and solvent immobilisation are favoured when conditions of random orientation prevail.

The apparent viscosity of a system which thins on shearing is most susceptible to changes in the shear rate in the intermediate range where there is a balance between randomness and alignment, and between aggregation and dispersion.

2. *Plasticity and yield value*—Plasticity is similar to shear-thinning, except that the system does not flow noticeably until the

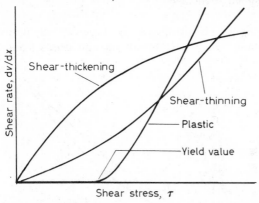

Figure 9.5. Steady-state forms of non-Newtonian flow

shearing stress exceeds a certain minimum value. The applied stress corresponding to a small but arbitrarily chosen rate of deformation is termed the *yield value*.

Plasticity is due to a continuous structural network which imparts rigidity to the sample and which must be broken before flow can occur. It is often difficult to distinguish between plastic and ordinary shear-thinning behaviour. Modelling clay, drilling muds and certain pigment dispersions are examples of plastic dispersions. Suspensions of carbon black in hydrocarbon oil often acquire a yield value on standing and become conducting owing to the contact between the carbon particles which is developed throughout the system.[139] Shearing reduces this conductivity, and the addition of deflocculating agents reduces both the conductivity and the yield value.

3.   *Shear-thickening*—Shear-thickening is characterised by an increase in apparent viscosity with increasing rate of deformation.

Shear-thickening is shown in particular, as a dilatant effect, by pastes of densely packed deflocculated particles in which there is only sufficient liquid to fill the voids. As the shear rate is increased, this dense packing must be broken down to permit the particles to flow past one another. The resulting expansion leaves insufficient liquid to fill the voids and is opposed by surface tension forces. This explains why wet sand apparently becomes dry and firm when walked upon.

Time-dependent phenomena

1.   *Thixotropy*—Thixotropy is the time-dependent analogue of shear-thinning and plastic behaviour and arises from somewhat

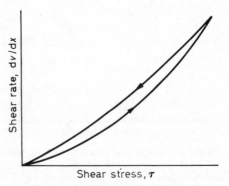

Figure 9.6. A thixotropic loop

similar causes. If a thixotropic system is allowed to stand and is then sheared at a constant rate, the apparent viscosity decreases with time until a balance between structural breakdown and structure re-formation is reached. If the sheared system is then allowed to stand, it eventually regains its original structure. A thixotropic hysteresis loop (Figure 9.6) can be obtained by measuring the non-equilibrium shear stress as the shear rate is first increased and then decreased in a standard way.

Solutions of high polymers are generally thixotropic to a certain extent; intermolecular attractions and entanglements are overcome and the extent of solvent immobilisation is reduced on shearing, whilst Brownian motion restores the system to its original condition when left to stand. The classical examples of thixotropic behaviour are given by the weak gel systems, such as flocculated sols of ferric oxide, alumina and many clays (particularly bentonite clays), which can be 'liquefied' on shaking and 'solidify' on standing. Thixotropy is particularly important in the paint industry, as it is desirable that the paint should flow only when being brushed on to the appropriate surface (high rate of shear) and immediately after brushing.

2.   *Rheopexy*—This is time-dependent shear-thickening, and is sometimes observed as an acceleration of thixotropic recovery— for example, bentonite clay suspensions often set only slowly on standing but quite rapidly when gently disturbed.

Irreversible phenomena

Shearing sometimes leads to an irreversible breaking *(rheodestruction)* of linkages between the structural elements of a material, e.g. with dehydrated silica gel networks.

Work hardening can occur as a result of mechanical entanglement or jamming of the structural elements on shearing, an example of this being the 'necking' and corresponding toughening of metal rods when subjected to a tensile stress. A technically important rheological property, which is related to strain hardening (and to flow elasticity), is *spinability*, i.e. the facility with which a material can be drawn into threads.

VISCOELASTICITY

When a typical elastic solid is stressed it immediately deforms by an amount proportional to the applied stress and maintains a constant deformation as long as the stress remains constant, i.e. it obeys Hooke's law. On removing the stress, the elastic energy stored in the solid is released and the solid immediately recovers its original shape. Newtonian liquids, on the other hand, deform at a rate proportional to the applied stress and show no recovery when the stress is removed, the energy involved having been dissipated as heat in overcoming the internal frictional resistance.

When viscoelastic materials are stressed, some of the energy involved is stored elastically, various parts of the system being deformed into new non-equilibrium positions relative to one another. The remainder is dissipated as heat, various parts of the system flowing into new equilibrium positions relative to one another. If the relative motion of the segments into non-equilibrium positions is hampered, the elastic deformation and recovery of the material is time-dependent *(retarded elasticity)*.

Experimental methods

Numerous instruments (plastometers, penetrometers, extensiometers, etc.) and procedures have been devised for measuring the

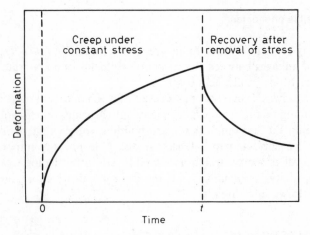

Figure 9.7. Creep and recovery curve for a typical viscoelastic material

rheological behaviour of various viscoelastic materials. However, the results obtained from most of these instruments are of little fundamental significance because the applied stress is not uniformly distributed throughout the sample, and the way in which the material behaves towards a particular apparatus is measured rather than a fundamental property of the material itself. Nevertheless, such empirical instruments are indispensable for control testing purposes in industry, where an arbitrary number which bears some relation to the mechanical property under consideration is usually quite sufficient.

To measure elastic and viscous properties which are characteristic of the material under consideration and independent of the nature of the apparatus employed, the applied stress and the resulting deformation must be uniform throughout the sample. Concentric cylinder and cone and plate methods approximate these requirements. For materials which are self-supporting, measurements on, for example, the shearing of rectangular samples are ideal.

Creep measurements involve the application of a constant stress (usually a shearing stress) to the sample and the measurement of the resulting sample deformation as a function of time. Figure 9.7 shows a typical creep and recovery curve. In *stress-relaxation* measurements, the sample is subjected to an instantaneous predetermined deformation and the decay of the stress within the sample as the structural segments flow into more relaxed positions is measured as a function of time.

The response of a material to an applied stress after very short times can be measured dynamically by applying a sinusoidally varying stress to the sample. A phase difference, which depends on the viscoelastic nature of the material, is set up between stress and strain. For Hookean elastic solids the stress and strain are in phase, whereas for purely viscous liquids the strain lags 90° behind the applied stress.

Time-dependent deformation and structural characteristics

The deformation of a material when subjected to a constant stress is, as discussed, generally time-dependent. At times of *ca.* 10^{-6} s and less all materials, including liquids, have *shear compliances* (i.e. shear/shear stress) of *ca.* $10^{-11} m^2 N^{-1}$ to $10^{-9} m^2 N^{-1}$. This is because there is only sufficient time available for an alteration of inter-atomic distances and bending of bond angles to take place, and the response of all materials is of the same order of magnitude in this respect. The time required for the various structural units of a material to move into new positions relative to one another depends on the size and shape of the units and the strength of the bonds between them.

The molecules of a liquid start to move relative to one another and the shear compliance increases rapidly after the shearing force has been applied for only *ca.* 10^{-6} s. On the other hand, hard solids, such as diamond, sodium chloride crystals and materials at a low enough temperature to be in the glassy state, show only the above rapid elastic deformation, even after the shearing stress has been applied for a considerable time.

The time taken for the structural units in viscoelastic materials, such as high polymers, to flow into new positions relative to one another is within a few decades of 1 second. Polymer molecules are in a continual state of flexing and twisting owing to their thermal energy. The configurations of the polymer chains alter more rapidly on a local scale than on a long-range scale. Under the influence of an external stress the polymer molecules flex and twist into more re-laxed positions, again more rapidly on a local scale. In general, there is a continuous range on the time scale covering the response of such systems to external stresses. On this basis, information concerning

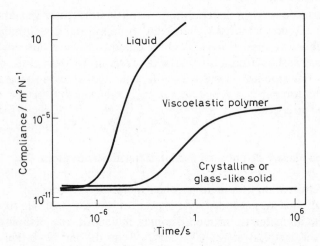

Figure 9.8. The time-dependence of deformations under constant stress

the structural nature of viscoelastic materials (particularly high polymers) can be obtained by measuring the compliance over a wide range on the time scale[140] (using dynamic methods for times of less than *ca.* 1 s, and creep measurements for times of greater than *ca.* 1 s).

Rubber elasticity[141]

Rubber-like materials *(elastomers)* have a structure based on polymer chains (e.g. polyisoprene chains

$$—CH_2—C(CH_3) = CH—CH_2—$$

in natural rubber) anchored at various points of cross-linkage. The extent of cross-linking can be increased by vulcanisation. When

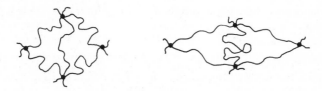

Figure 9.9. The stretching of rubber

stretched, the polymer chains are extended lengthwise and compressed crosswise (Figure 9.9) from their average configurations, the change in total volume being inappreciable, and thermal restoring forces are generated. When the tension is released the polymer chains return (usually rapidly) by thermal motion to their original average configurations.

Owing to the thermal nature of the restoring force, the deformation of rubber, for a given load, decreases as the temperature is raised. This contrasts with the elasticity of a metal spring, which is due to individual atoms being slightly displaced from their local equilibrium positions, the coil structure greatly multiplying this effect, and which increases with increasing temperature.

If the degree of cross-linking is not very great, as in crude rubber, viscous flow can occur, the polymer chains moving permanently into new equilibrium positions. Excessive cross-linking, on the other hand, restricts changes in the chain configurations and the rubber becomes hard and difficult to deform.

Partial crystallisation may take place in polymeric materials especially when stretched and/or cooled. From the mechanical standpoint, the introduction of crystalline regions in a polymer is equivalent to increasing the degree of cross-linking, and a partial loss of elasticity results.

Polymers exhibit a glass transition temperature below which the chain arrangements are frozen. Thermal motion no longer over-comes the attractive forces between the polymer chains, and the sample becomes hard and brittle.

Non-linear viscoelasticity

Viscoelasticity is termed *linear* when the time-dependent compliance (strain/stress) of a material is independent of the magnitude of the applied stress. All materials have a linearity limit (see Table 9.2).

Table 9.2   LINEAR VISCOELASTICITY LIMITS

Material	Stress/N m^{-2}	Percentage strain
Elastomers	ca. 10^6 to 10^7	ca. 10 to 100
Plastics	ca. 10^6 to 10^7	ca. 0·1 to 1
Fats	ca. 10^2	ca. 0·01

The linearity limit of elastomers is large, because their deformation is of an entropic nature and does not involve bond rupture and reformation.

Viscoelastic materials have much lower linearity limits. For the segments or particles in such systems to move (flow) relative to one another without weakening the material, the forces (specific and non-specific) between them must be overcome and then reinstated at the same rate in new positions. If the deforming stress is such that these forces are not reinstated as rapidly as they are overcome, the material becomes structurally weaker. The remaining forces in certain cross-sections between the structural units are then over-come even more readily by the applied stress, and cracks might appear in the sample. Materials with low linear viscoelastic limits are, therefore, those which are readily *work softened.*

The Weissenberg effect

A characteristic of viscoelastic behaviour is the tendency for flow to occur at right angles to the applied force. An extreme example of this behaviour is illustrated in Figure 9.10. When a rotating rod is lowered into a Newtonian liquid the liquid is set into rotation and tends to move outwards, leaving a depression around the rod. When the rotating rod is lowered into a viscoelastic liquid, the liquid may actually climb up the rod. The rotation of the rod causes the

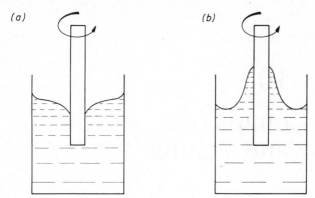

Figure 9.10. The Weissenberg effect: (*a*) Newtonian liquid; (*b*) viscoelastic liquid

liquid to be sheared circularly and, because of its elastic nature, it acts like a stretched rubber band tending to squeeze liquid in towards the centre of the vessel and, therefore, up the rod.

10
Emulsions
and Foams

OIL IN WATER AND WATER IN OIL EMULSIONS[142, 143]

An emulsion is a dispersed system in which the phases are immiscible
or partially miscible liquids. The globules of the dispersed liquid
are generally between $0 \cdot 1\,\mu m$ and $10\,\mu m$ in diameter, and so are
larger than the particles found in sols.

In nearly all emulsions, one of the phases is aqueous and the other
is (in the widest sense of the term) an oil. If the oil is the dispersed
phase, the emulsion is termed an *oil in water* (O/W) emulsion; if
the aqueous medium is the dispersed phase, the emulsion is termed
a *water in oil* (W/O) emulsion. There are several methods by which
the emulsion type may be identified.

1. Generally, an O/W emulsion has a creamy texture and a W/O
   emulsion feels greasy.
2. The emulsion mixes readily with a liquid which is miscible
   with its dispersion medium.
3. The emulsion is readily coloured by dyes which are soluble
   in the dispersion medium.
4. O/W emulsions generally have a much higher electrical conductivity than W/O emulsions.

Emulsifying agents and emulsion stability

Probably the most important physical property of an emulsion is its stability. The term 'emulsion stability' can be used with reference to three essentially different phenomena—creaming (or sedimentation), flocculation and a breaking of the emulsion due to droplet coalescence.

Creaming results from a density difference between the two phases and is not necessarily accompanied by droplet flocculation, although this facilitates the process.

Droplet collisions may result in flocculation, which in turn may lead to coalescence into larger globules. Eventually, the dispersed phase may become a continuous phase, separated from the dispersion medium by a single interface. The time taken for such phase separation may be anything from seconds to years, depending on the emulsion formulation and manufacturing conditions.

If an emulsion is prepared by homogenising two pure liquid components, phase separation will generally be rapid, especially if the concentration of the dispersed phase is at all high. To prepare reasonably stable emulsions a third component—an *emulsifying agent* (or *emulsifier*)—must be present. The materials which are most effective as emulsifying (and foaming) agents can be broadly classified as:

1. Surface-active materials
2. Naturally occurring materials
3. Finely divided solids

The functions of the emulsifying agent are to facilitate emulsification and promote emulsion stability. The emulsifying agent forms an adsorbed film around the dispersed droplets which helps to prevent flocculation and coalescence. The stabilising mechanism is generally complex and may vary from system to system. In general, however, the factors which control droplet flocculation are the same as those which control the stability of sols (see Chapter 8), whereas stability against droplet coalescence depends mainly on the mechanical properties of the interfacial film. The following factors (which depend on the nature of the emulsifying agent and/or on a suitable choice of formulation and manufacturing conditions) favour emulsion stability.

1. *Low interfacial tension*—The adsorption of surfactant at oil–

water interfaces causes a lowering of interfacial energy, thus facilitating the development and enhancing the stability of the large interfacial areas associated with emulsions.

2. *A mechanically strong interfacial film*—The stability of emulsions stabilised by proteins arises from the mechanical protection given by the adsorbed films around the droplets, rather than from a reduction of interfacial tension.

Finely divided solids for which the contact angle is between $0°$ and $180°$ have a tendency to collect at the oil–water interface (cf. flotation, page 122), where they impart stability to the emulsion.

Surfactants can also stabilise in the mechanical sense. Coalescence involves droplet flocculation followed by a squeezing of film material from the region of droplet contact,[144] and the latter is more favoured with an expanded film than with a close-packed film. For example, very stable hydrocarbon oil in water emulsions can be prepared with sodium cetyl sulphate (dissolved in the water) plus cetyl alcohol (dissolved in the oil) as emulsifier (a condensed mixed film being formed at the interface), whereas hydrocarbon oil in water emulsions prepared with sodium cetyl sulphate plus oleyl alcohol (which gives an expanded mixed film) are much less stable.[145]

3. *Electrical double layer repulsions* (see page 169)

4. *Relatively small volume of dispersed phase*

5. *Small droplet size*

6. *High viscosity*—A high Newtonian viscosity simply retards the rates of creaming, coalescence, etc. If a weak gel network is formed, for example, by dissolving sodium carboxymethyl cellulose in an O/W emulsion, genuine stability might ensue. However, the overall rheological properties of such an emulsion may not be acceptable.

Emulsifying agents and emulsion type

The type of emulsion which is formed when a given pair of immiscible liquids is homogenised depends on (1) the relative volumes of the two phases, and (2) the nature of the emulsifying agent.

1. *Phase volume*—The higher its phase volume, the more likely

a liquid is to become the dispersion medium. However, the liquid with the greater phase volume need not necessarily be the dispersion medium.

If the emulsion consisted of an assembly of closely packed uniform spherical droplets, the dispersed phase would occupy 0.74 of the total volume. Stable emulsions can, however, be prepared in which the volume fraction of the dispersed phase exceeds 0.74, because (a) the droplets are not of uniform size and can, therefore, be packed more densely, and (b) the droplets may be deformed into polyhedra, the interfacial film preventing coalescence.

2. *Nature of the emulsifying agent*—Alkali-metal soaps favour the formation of O/W emulsions, whereas heavy-metal soaps favour

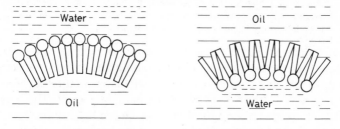

Figure 10.1. The oriented wedge theory

the formation of W/O emulsions. O/W emulsions in the middle concentration region stabilised by alkali-metal soaps can often be broken, and even inverted into W/O emulsions, by the addition of heavy-metal ions.

Several theories relating to emulsion type have been proposed.

The 'oriented wedge' theory[10] requires that (to achieve maximum interfacial density of emulsifier) the end of a surfactant emulsifier molecule which has the greater cross-sectional area should be oriented towards the dispersion medium (Figure 10.1). Thus monovalent soaps should tend to give O/W emulsions and polyvalent soaps should tend to give W/O emulsions. However, the theory is not entirely consistent with experimental evidence, since some monovalent soaps (e.g. silver soaps) tend to give W/O emulsions.

Another theory regards the film as duplex in nature with inner and outer interfacial tensions, the type of emulsion formed being such that the inner interfacial tension is the greater.

The most satisfactory general theory of emulsion type is that originally proposed for emulsions stabilised by finely divided solids (see Figure 10.2). If the solid is preferentially wetted by one of the phases, then more particles can be accommodated at the interface if the interface is convex towards that phase (i.e. if the preferentially wetting phase is the dispersion medium). For example, bentonite

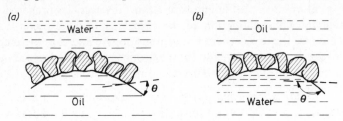

Figure 10.2. Stabilisation of emulsions by finely divided solids: (*a*) preferential wetting by water leading to an O/W emulsion; (*b*) preferential wetting by oil leading to a W/O emulsion

clays (which are preferentially wetted by water) tend to give O/W emulsions, whereas carbon black (which is preferentially oil wetted) tends to give W/O emulsions. This preferential wetting theory can be extended to cover other types of emulsifying agent. The type of emulsion which tends to form depends on the balance between the hydrophilic and lipophilic properties of the emulsifier—alkali-metal soaps favour the formation of O/W emulsions because they are more hydrophilic than lipophilic, whereas the reverse holds for heavy-metal soaps.

The amphiphilic nature of many emulsifying agents (particularly non-ionic surfactants) can be expressed in terms of an empirical scale of so-called HLB (hydrophile-lipophile balance) numbers[146]

Table 10.1   HLB VALUES

Applications	Dispersibility in water
3–6 W/O emulsions	1–4 Nil
7–9 Wetting agents	3–6 Poor
8–15 O/W emulsions	6–8 Unstable milky dispersion
13–15 Detergent	8–10 Stable milky dispersion
15–18 Solubiliser	10–13 Translucent dispersion/solution
	13– Clear solution

(see Table 10.1). The least hydrophilic surfactants are assigned the lowest HLB values. A number of different formulae have been established for calculating HLB numbers from composition data and they can also be determined experimentally, e.g. from cloud-point measurements.[142, 143] For mixed emulsifiers, approximate algebraic additivity holds.

The optimum HLB number for forming an emulsion depends to some extent on the nature of the particular system. Suppose that 20 per cent sorbitan tristearate (HLB 2·1)+80 per cent polyoxy-ethylene sorbitan monostearate (HLB 14·9) is the optimum composition of a mixture of these emulsifiers for preparing a particular O/W emulsion. The HLB of the mixture is, therefore, $(0·2 \times 2·1) + (0·8 \times 14·9) = 12·3$. The theory is that an HLB of 12·3 should be optimum for the formation of this particular O/W emulsion using other emulsifier systems; for example, the optimum proportions in a mixture of sorbitan mono-oleate (HLB 4·3) and polyoxyethylene sorbitan monopalmitate (HLB15·6) should be approximately 30 per cent and 70 per cent, respectively. In commercial emulsion formulation, HLB numbers are used advantageously in this way as an initial guide prior to a certain amount of trial and error testing.

Breaking of emulsions

In many instances it is the breaking of an emulsion (demulsification) which is of practical importance. Examples are the creaming, breaking and inversion of milk to obtain butter, and the breaking of W/O oil-field emulsions.

A number of techniques are used commercially to accelerate emulsion breakdown. Mechanical methods include centrifugal separation, freezing, distillation and filtration. Another method is based on the principle of antagonistic action, i.e. the addition of O/W-promoting emulsifiers tends to break W/O emulsions, and vice versa. Emulsions can also be broken by the application of intense electrical fields, the principal factors involved being electrophoresis in the case of O/W emulsions and droplet deformation in the case of W/O emulsions.

EMULSION POLYMERISATION

An interesting application of emulsion technique is to be found in the process of emulsion polymerisation. Unless the degree of polymerisation is sharply limited, considerable processing difficulties are experienced in bulk polymerisation. Such difficulties arise mainly from the exothermic nature of polymerisation reactions and the necessity for efficient cooling to avoid the undesirable effects associated with a high reaction temperature (see page 13). Even at moderate degrees of polymerisation the resulting high viscosity of the reaction mixture makes stirring and efficient heat transfer very difficult.

The difficulties associated with heat transfer can be overcome, and higher molecular weight polymers obtained, by the use of an emulsion system. The heat of polymerisation is readily dissipated into the aqueous phase and the viscosity of the system changes only slightly during the reaction. A typical recipe[6] for the polymerisation of vinyl monomers would be to form an O/W emulsion from:

Monomer	100 g
Emulsifying agent (fatty-acid soap)	2 g to 5 g
Catalyst (potassium persulphate)	0·1 g to 0·5 g
Water	180 g

The basic theory of emulsion polymerisation is due to Harkins.[10] Monomer is distributed throughout the emulsion system (1) as stabilised emulsion droplets, (2) dissolved to a small extent in the aqueous phase where initiation can take place, and (3) solubilised in soap micelles. Polymerisation does not take place in the emulsified monomer droplets but in the soap micelles. The monomer droplets serve as reservoirs to supply material to the polymerisation sites by diffusion through the aqueous phase. As the micelles grow, they adsorb free emulsifier from solution, and eventually from the surface of the emulsion droplets. The emulsifier thus serves to stabilise the polymer particles. This theory accounts for the observation that the rate of polymerisation and the number of polymer particles finally produced depend largely on the emulsifier concentration, and that the number of polymer particles may far exceed the number of monomer droplets initially present.

Monodispersed sols containing spherical polymer particles (e.g. polystyrene lattices[147]) can be prepared by emulsion polymerisation, and are particularly useful as model systems for studying various aspects of colloidal behaviour. The seed sol (see page 11) is prepared with the emulsifier concentration well above the c.m.c., then, with the emulsifier concentration below the c.m.c., subsequent growth of the seed particles is achieved without the formation of further new particles.

FOAMS[148, 149]

A foam is a coarse dispersion of gas in liquid, and two extreme structural situations can be recognised. The first type (dilute foams) consist of nearly spherical bubbles separated by rather thick films of somewhat viscous liquid. The other type (concentrated foams) are mostly gas phase, and consist of polyhedral gas cells separated by thin liquid films (which may develop from more dilute foams as a result of liquid drainage, or directly from a liquid of relatively low viscosity).* The nature of thin liquid films (as found in these concentrated foams) is currently the subject of a great deal of fundamental research.

Foam stability

Only transitory foams can be formed with pure liquids and, as with emulsions, a third (surface active) component—a *foaming agent*—is necessary to achieve any reasonable degree of stability. Good emulsifying agents are, in general, also good foaming agents, since the factors which influence emulsion stability (against droplet coalescence) and foam stability (against bubble collapse) are somewhat similar.

The stability of a foam depends upon two principal factors—the tendency for the liquid films to drain and become thinner, and their tendency to rupture as a result of random disturbances.[148, 149]

Owing to their high interfacial area (and surface free energy), all foams are unstable in the thermodynamic sense. Some distinction can be made, however, between *unstable* and *metastable* foam structures. *Unstable* foams are typified by those formed from aqueous solutions of short chain fatty acids or alcohols. The presence of these mildly surface-active agents retards drainage and film rupture to some extent, but does not stop these processes from continuously

* Similarly, some solid foams (e.g. foam rubber) consist of spherical gas bubbles trapped within a solid network, whereas others (e.g. expanded polystyrene) consist of as little as 1 per cent solid volume and are composed of polyhedral gas cells separated by very thin solid walls.

taking place to the point of complete foam collapse. *Metastable* foams are typified by those formed from solutions of soaps, synthetic detergents, proteins, saponins, etc. The balance of forces is such that the drainage of liquid stops when a certain film thickness is reached and, in the absence of disturbing influences (such as vibration, draughts, evaporation, diffusion of gas from small bubbles to large bubbles, heat, temperature gradients, dust and other impurities), these foams would persist almost indefinitely.

Foam drainage

Consider, as a simple representation of the liquid in a foam system, the drainage of a single vertical liquid film. Suppose that such a film is formed by carefully raising a rectangular frame from a soap solution in a saturated atmosphere to avoid evaporation. Initially the film will be relatively thick and drainage will take place mainly by gravitational flow of liquid throughout the whole of the film. When the film has attained a thickness of the order of micrometres, gravitational flow down the laminar part of the film will have become extremely slow (even if the liquid has a low viscosity) and the predominant drainage mechanism will involve liquid being discharged from the central laminar region and flowing down the relatively wide liquid columns (known as Plateau borders) adjacent to the supporting frame.

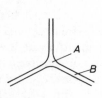

Figure 10.3. Plateau border at a line of intersection of three bubbles. Owing to the curvature of the liquid–gas interface at *A*, the pressure of liquid at *A* is lower than that at *B*, thus causing capillary flow of liquid towards *A*

As a result of drainage, the film thickness will be greater at the bottom than at the top and spectral colours will develop as a result of interference between light reflected from the front and back surfaces of the film. As film drainage proceeds these coloured bands will move downwards and the spacings between them increase until eventually a silvery and then a black film develops which is thin enough for light reflected from the front and back surfaces to interfere constructively for all visible wavelengths. Black soap films of about 5 nm thickness (which is not much greater than the length of two soap molecules) have been formed and studied.

The discharge of liquid from the laminar part of a thin film is governed by the pressure of the liquid in this region compared with that of the liquid in the Plateau borders. At least three factors are likely to be involved. van der Waals attractive forces favour film

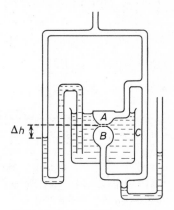

Figure 10.4. Apparatus for measuring the disjoining pressure of free films as a function of their thickness[150] (After B. V. Derjaguin and A. S. Titijevskaya)

thinning, and the overlapping of similarly charged electric double layers opposes film thinning (see Chapter 8). The other important factor is a capillary pressure, which favours film thinning. This arises because the pressure of the adjacent gas phase is uniform and, therefore, the pressure of liquid in the Plateau borders, where the interface is curved, is less than that of liquid in the laminar film. Depending on the balance of these forces, a film may either thin continuously and eventually rupture, or attain an equilibrium thickness. Any propagated structure within the film may significantly affect the equilibrium film thickness that a balance between the above forces would otherwise determine.

Experimental studies on non-draining, horizontal liquid films in which the equilibrium film thickness is measured as a function of ionic strength and applied hydrostatic pressure (or suction) provides a means of investigating the above forces.[15, 150, 151, 152] Figure 10.4 shows an apparatus used by Derjaguin and Titijevskaya.[150]

A flat liquid film with an area of *ca.* 1 mm² is formed between cups *A* and *B* which are connected via tube *C* to equalise bubble pressures. The pressure in the bubbles in excess of that in the liquid film is calculated from the manometer reading Δh. Derjaguin calls this excess pressure the *disjoining pressure*. An elaborate optical device (not shown) permits measurement of the film thickness.

Figure 10.5 shows the results of some measurements on aqueous

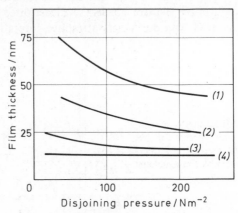

Figure 10.5. Film thickness as a function of disjoining pressure for films of 10^{-3} mol dm⁻³ aqueous sodium oleate containing NaCl at concentrations of (*1*) 10^{-4} mol dm⁻³; (*2*) 10^{-3} mol dm⁻³; (*3*) 10^{-2} mol dm⁻³; and (*4*) 10^{-1} mol dm⁻³ (After B. V. Derjaguin and A. S. Titijevskaya[150])

sodium oleate films. The sensitivity of the equilibrium film thickness to added electrolyte reflects qualitatively the expected positive contribution of electric double layer repulsion to the disjoining pressure. However, this sensitivity to added electrolyte is much less than that predicted from electric double layer theory and at high electrolyte concentration an equilibrium film thickness of *ca.* 12 nm is attained which is almost independent of the magnitude of the disjoining pressure. To account for this observation, Derjaguin and Titijevskaya have postulated the existence of hydration layers effectively *ca.* 6 nm thick. Similar conclusions have also been reached by van den Tempel[153] from experiments with oil droplets in aqueous sodium dodecyl sulphate plus sodium chloride.

Film rupture

In addition to film drainage, the stability of a foam depends on the ability of the liquid films to resist excessive local thinning and rupture which may occur as a result of various random disturbances. A number of factors may be involved with varying degrees of importance depending on the nature of the particular foam in question.

1. *Gibbs-Marangoni surface elasticity effect*—This is an important stabilising effect in foams which are formed from solutions of soaps, detergents, etc. If a film is subjected to local stretching as a result of some external disturbance, the consequent increase in surface area will be accompanied by a decrease in the surface excess concentration of foaming agent and, therefore, a local increase in surface tension (Gibbs effect). Since a certain time is required for surfactant molecules to diffuse to this surface region and restore the original surface tension (Marangoni effect), this increased surface tension may persist for long enough to cause the disturbed film region to recover its original thickness.

As an extension of the Marangoni effect, Ewers and Sutherland[154] have suggested a surface transport mechanism in which the surface tension gradient created by local film thinning causes foaming agent to spread along the surface and drag with it a significant amount of underlying solution, thus opposing the thinning process.

An absence of the Gibbs-Marangoni effect is the main reason why pure liquids do not foam. It is also interesting, in this respect, to observe that foams from moderately concentrated solutions of soaps, detergents, etc., tend to be less stable than those formed from more dilute solutions. With the more concentrated solutions, the increase in surface tension which results from local thinning is more rapidly nullified by diffusion of surfactant from the bulk solution. The opposition to fluctuations in film thickness by corresponding fluctuations in surface tension is, therefore, less effective.

2. *Surface rheology*—The mechanical properties of the surface films (as in the case of emulsions) often have a considerable influence on foam stability. Several considerations may be involved.

A high bulk liquid viscosity simply retards the rate of foam collapse. High surface viscosity, however, involves strong retardation of bulk liquid flow close to the surfaces and, consequently, the drainage of thick films is considerably more rapid than that of thin films, thus facilitating the attainment of a uniform film thickness.

Surface elasticity facilitates the maintenance of a uniform film thickness, as discussed above; however, the existence of rigid, condensed surface films is detrimental to foam stability owing to the very small changes in area over which such films show elasticity.

3.   *Equilibrium film thickness*—If the balance of van der Waals attraction, electric double layer repulsion, capillary pressure, structure propagation, etc., favours an equilibrium film thickness, random fluctuations in film thickness will, in any case, tend to be neutralised.

Antifoaming agents

The prevention of foaming and the destruction of existing foams is often a matter of practical importance; for example, polyamides and silicones find use as foam inhibitors in water boilers. Antifoaming agents act against the various factors which promote foam stability (described above) and, therefore, a number of mechanisms may be operative.

Foam inhibitors are, in general, materials which tend to be adsorbed in preference to the foaming agent, yet do not have the requisites to form a stable foam. They may be effective by virtue of rapid adsorption; for example, the addition of tributyl phosphate to aqueous sodium oleate solutions significantly reduces the time required to reach equilibrium surface tension,[155] thus lessening the Marangoni surface elasticity effect and the foam stability. They may also act, for example, by reducing electric double layer repulsion or by facilitating drainage by reducing hydrogen bonding between the surface films and the underlying solution.

Foams can often be broken by spraying with small quantities of substances such as ether and n-octanol. As a result of their high surface activity, these foam breakers raise the surface pressure over small regions of the liquid films and spread from these regions displacing the foaming agent and carrying with them some of the underlying liquid.[154] Small regions of film are, therefore, thinned and left without the properties to resist rupture.

Problems

$$k = 1.380\,5 \times 10^{-23}\,\text{J}\,\text{K}^{-1}$$
$$N_A = 6.022\,5 \times 10^{23}\,\text{mol}^{-1}$$
$$R = 8.314\,3\,\text{J}\,\text{K}^{-1}\,\text{mol}^{-1}$$
$$e = 1.602\,1 \times 10^{-19}\,\text{C}$$
$$\varepsilon_0 = 8.854\,2 \times 10^{-12}\,\text{kg}^{-1}\,\text{m}^{-3}\,\text{s}^4\,\text{A}^2$$
$$g = 9.806\,6\,\text{m}\,\text{s}^{-2}$$

Volume of ideal gas at s.t.p. (0°C and 1 atm) = $2.241\,4 \times 10^{-2}\,\text{m}^3$ mol^{-1}

1 atm = 760 Torr = $1.013\,25 \times 10^5\,\text{N}\,\text{m}^{-2}$

0°C = 273.15K

ln 10 = 2.302 6

π = 3.141 56

1.   Calculate the average displacement in 1 minute along a given axis produced by Brownian motion for a spherical particle of radius $0.1\,\mu m$ suspended in water at 25°C. The coefficient of viscosity of water at this temperature is $8.9 \times 10^{-4}\,\text{kg}\,\text{m}^{-1}\,\text{s}^{-1}$.

2.   The sedimentation and diffusion coefficients for myoglobin in dilute aqueous solution at 20°C are $2.04 \times 10^{-13}\,\text{s}$ and $1.13 \times 10^{-10}\,\text{m}^2\,\text{s}^{-1}$, respectively. The partial specific volume of the protein is $0.741\,\text{cm}^3\,\text{g}^{-1}$, the density of the solution is $1.00\,\text{g}\,\text{cm}^{-3}$ and the coefficient of viscosity of the solution is $1.00 \times 10^{-3}\,\text{kg}\,\text{m}^{-1}\,\text{s}^{-1}$. Calculate (a) the relative molecular mass, and (b) the frictional ratio

of this protein. What is the probable shape of a dissolved myoglobin molecule?

3. An aqueous solution of β-lactoglobulin in the presence of sufficient electrolyte to eliminate charge effects was centrifuged to equilibrium at 11 000 revolutions per minute and at 25°C. The following equilibrium concentrations were measured:

Distance from axis of rotation/cm	4·90	4·95	5·00	5·05	5·10	5·15
Concentration/g dm^{-3}	1·30	1·46	1·64	1·84	2·06	2·31

The partial specific volume of the protein was $0·75 \text{ cm}^3 \text{ g}^{-1}$ and the density of the solution (assumed constant) was $1·0 \text{ g cm}^{-3}$. Calculate the relative molecular mass of the protein.

4. The following osmotic pressures were measured for solutions of a sample of polyosobutylene in benzene at 25°C:

Concentration/(g/100 cm^3)	0·5	1·0	1·5	2·0
Osmotic pressure/(cm solution)	1.03	2·10	3·22	4·39

(solution density $= 0·88 \text{ g cm}^{-3}$ in each case)

Calculate an average relative molecular mass.

5. A $0·01 \text{ mol dm}^{-3}$ aqueous solution of a colloidal electrolyte (which can be represented as $Na_{15}X$) is on one side of a dialysis membrane, whilst on the other side of the membrane is an equal volume of $0·05 \text{ mol dm}^{-3}$ aqueous sodium chloride. When Donnan equilibrium is established, what net fraction of the NaCl will have diffused into the compartment containing the colloidal electrolyte?

6. The following light scattering data give values of the quantity $10^7 \quad c/R_\theta$ for solutions of cellulose nitrate dissolved in acetone:

Concentration/ g dm^{-3}	Scattering angle (in relation to transmitted beam)		
	45°	32°	17°30'
0·88	69·8	49·0	33·0
0·64	66·0	45·5	29·4
0·43	62·8	42·1	25·9

Make appropriate extrapolations and calculate an average relative molecular mass for the cellulose nitrate.

7. If bubbles of air 50 nm in diameter and no other nuclei are present in water just below boiling point, approximately how much could water be superheated at normal atmospheric pressure before boiling starts? The surface tension of water at 100°C is 59 mN m^{-1} and its latent heat of vaporisation is 2·25 kJ g^{-1}.

8. The following surface tensions were measured for aqueous solutions of sodium dodecyl sulphate at 25°C:

$c/10^{-3}$ mol dm^{-3}	0	1·0	2·0	3·0	4·0	5·0	6·0	7·0	8·0
γ/mN m^{-1}	72·7	67·9	62·3	56·7	52·5	48·8	45·6	42·8	40·5

Calculate the surface pressure of the adsorbed film, the surface excess concentration and the average area occupied by each adsorbed molecule for bulk concentrations (in 10^{-3} mol dm^{-3}) of 2·0, 4·0 and 6·0.

9. At 20°C the surface tensions of water and mercury are 73 mN m^{-1} and 485 mN m^{-1}, respectively, and the interfacial tension of the mercury–water interface is 375 mN m^{-1}. Calculate:
   (a) the work of adhesion between mercury and water,
   (b) the work of cohesion for (i) mercury, and (ii) water,
   (c) the initial spreading coefficient of water on mercury.

10. The contact angle for water on paraffin wax is 105° at 20°C. Calculate the work of adhesion and the spreading coefficient. The surface tension of water at 20°C is 72·75 mN m^{-1}.

11. The pressure required to prevent liquid from entering a plug of a finely divided solid is twice as great for a liquid of surface tension 50 mN m^{-1}, which completely wets the solid, as it is for a liquid of surface tension 70 mN m^{-1}, which has a finite contact angle with the solid. Calculate this contact angle.

12. The following surface pressure measurements were obtained for a film of haemoglobin spread on 0·01 mol dm^{-3} HCl aq. at 25°C:

A/m^2 mg^{-1}	4·0	5·0	6·0	7·5	10·0
π/mN m^{-1}	0·28	0·16	0·105	0·06	0·035

Calculate the relative molecular mass of the spread protein and compare your answer with a value of 68 000 as determined from sedimentation measurements.

13. The following data refer to the adsorption of hydrogen on a nickel catalyst at 25°C:

Pressure/
$10^5 \, N \, m^{-2}$ 0·005 0·010 0·015 0·02 0·03 0·05 0·10 0·15
Volume of gas
adsorbed/cm³
(reduced to s.t.p.) 0·48   0·68   0·83 0·93 1·07 1·20 1·33 1·40

Show that the data fit a Langmuir adsorption isotherm expression and evaluate the constants.

14. The following results were obtained for the adsorption of nitrogen on rutile at 77K:

Relative pres-
sure (p/p_0)   0·02 0·05 0·1 0·2   0·3   0·4   0·6   0·8   0·9   0·95
Volume of gas
adsorbed/cm³
(s.t.p.) g^{-1} 2·1 2·7 3·4 4·2   4·8   5·3   6·7   8·0 11·8 20·5

Plot the adsorption isotherm and use the BET equation to calculate a specific surface area for the rutile sample, taking the molecular area of nitrogen to be $16·2 \times 10^{-20} \, m^2$.

15. The following data refer to the adsorption of n-butane at 273K by a sample of tungsten powder which has a specific surface area (as determined from nitrogen adsorption measurements at 77K) of 6·5 m² g⁻¹:

Relative pressure (p/p_0)	0·04	0·10	0·16	0·25	0·30	0·37
Volume of gas adsorbed/ cm³ (s.t.p.) g^{-1}	0·33	0·46	0·54	0·64	0·70	0·77

Use the BET equation to calculate a molecular area for the adsorbed butane at monolayer coverage and compare it with the value of $32·1 \times 10^{-20} \, m^2$ estimated from the density of liquid butane.

16. The following data refer to the adsorption of nitrogen on a sample of carbon black at 77K:

Relative pressure (p/p_0)	0·05	0·10	0·15	0·20	0·25	0·30
Volume of N_2 adsorbed/ cm³ (s.t.p.) g^{-1}	12·4	14·5	16·2	17·6	19·0	20·5

Use the BET equation to calculate a specific surface area and compare it with $42\,m^2\,g^{-1}$ as determined by means of an electron microscope. The molecular area of nitrogen is $16\cdot2\times10^{-20}\,m^2$.

17. The following results refer to the adsorption of nitrogen on a graphitised sample of carbon black, and give the ratio of the nitrogen pressures for temperatures of 90K and 77K which are required to achieve a given amount of adsorption:

Amount of N_2 adsorbed (V/V_m)	0·4	0·8	1·2
p_{90K}/p_{77K}	14·3	17·4	7·8
	15.1	18.5	8.3

Calculate an isosteric heat of adsorption for each value of V/V_m and comment on the values obtained.

18. Use the Kelvin equation to calculate the pore radius which corresponds to capillary condensation of nitrogen at 77K and a relative pressure of 0·75. Allow for multilayer adsorption on the pore wall by taking the thickness of the adsorbed layer on a nonporous solid as 0·9 nm at this relative pressure. List the assumptions upon which this calculation is based. For nitrogen at 77K, the surface tension is $8\cdot85\,mN\,m^{-1}$ and the molar volume is $34\cdot7\,cm^3\,mol^{-1}$.

19. The following data refer to the adsorption of dodecanol from solution in toluene by a sample of carbon black, the BET nitrogen adsorption area of which is $105\,m^2\,g^{-1}$:

Equilibrium concentration/ mol dm^{-3}: 5×10^{-4} 6.94×10^{-4} 8.9×10^{-4} 1.3×10^{-3} 1.63×10^{-3}

Equilibrium concentration/ mol dm^{-3}	0·012	0·035	0·062	0·105	0·148
Amount adsorbed/μmol g^{-1}	24·1	50·4	69·8	81·6	90·7

Show that the data fit a Langmuir adsorption isotherm equation and calculate the area occupied by each adsorbed dodecanol molecule at limiting adsorption.

20. Spherical particles of radius $0\cdot4\,\mu m$ suspended in $0\cdot01\,mol\,dm^{-3}$ aqueous sodium chloride are observed to have an electrophoretic mobility of $2\cdot5\times10^{-8}\,m^2\,s^{-1}\,V^{-1}$ at 25°C. Calculate an approximate value for the zeta potential. In what sense will the simplifications upon which your calculation is based affect the answer? At 25°C, the dielectric constant of water is 78·5 and the coefficient of viscosity of water is $8\cdot9\times10^{-4}\,kg\,m^{-1}\,s^{-1}$.

21. In a microelectrophoresis experiment a spherical particle of diameter $0\cdot5\,\mu m$ dispersed in a $0\cdot1\,mol\,dm^{-3}$ aqueous solution

of KCl at 25°C takes 8·0 s to cover a distance of 120 μm along one of the 'stationary levels' of the cell, the potential gradient being 10·0 V cm^{-1}. Calculate

(a) the electrophoretic mobility of the particle,
(b) the probable error in this single mobility determination arising from the Brownian motion of the particle during the course of the measurement,
(c) an approximate value for the zeta potential of the particle,
(d) an approximate value for the charge density at the surface of shear.

At 25°C the dielectric constant of water is 78·5 and the coefficient of viscosity of water is $8·9 \times 10^{-4}$ kg m^{-1} s^{-1}.

22. Calculate the rate of electro-osmotic flow of water at 25°C through a glass capillary tube 10 cm long and 1 mm diameter when the potential difference between the ends is 200 V. The zeta potential for the glass–water interface is -40 mV. The coefficient of viscosity and dielectric constant of water at 25°C are $8·9 \times 10^{-4}$ kg m^{-1} s^{-1} and 78·5, respectively.

23. Electrokinetic measurements at 25°C on silver iodide in 10^{-3} mol dm^{-3} aqueous potassium nitrate give $d\zeta/d(pAg) = -35$ mV at the zero point of charge. Assuming no specific adsorption of K^+ or NO_3^- ions and no potential drop within the solid, estimate the capacity of the inner part of the electric double layer. Taking the thickness of the inner part of the double layer to be 0·4 nm, what value for the dielectric constant near to the interface does this imply? Comment on the result.

24. The following results were obtained by particle counting during the flocculation of a hydrosol at 25°C by excess 1–1 electrolyte:

Time/min	0	2	3	5	8	11	15
Particle concentration/ 10^8 cm^{-3}	114	10·6	7·1	4·4	2·8	2·0	1·5

Calculate a 'second order' rate constant k_2 and compare it with the value k_2^0 calculated on the assumption that flocculation is a diffusion controlled process. The coefficient of viscosity of water at 25°C is $8·9 \times 10^{-4}$ kg m^{-1} s^{-1}.

25. The flow times in an Ostwald viscometer for solutions of polystyrene in toluene at 25°C are as follows:

Concentration/(g/100 cm^3)	0	0·4	0·8	1·2
Flow time/s	31·7	38·3	45·0	51·9

K and α in the expression, $[\eta] = KM_r^\alpha$, are 3.7×10^{-5} m^3 kg^{-1} and 0·62, respectively, for this polymer–solvent system. Assuming a constant density for the solutions, calculate an average relative molecular mass for the polystyrene sample. How would this relative molecular mass be expected to compare with the relative molecular mass of the same sample of polystyrene in toluene determined from (a) osmotic pressure, and (b) light scattering measurements?

Answers

1. $17 \cdot 2 \, \mu m$

2. $M_r = 17\,000$ (i.e. $M = 17 \cdot 0$ kg mol^{-1})
    $f/f_0 = 1 \cdot 11$ [which, allowing for hydration, suggests that dissolved myoglobin molecules are approximately spherical (see Figure 2.1)]

3. $M_r = 34\,000$ (i.e. $M = 34 \cdot 0$ kg mol^{-1}) (monodispersed)

4. $M_r = 143\,000$ (i.e. $M = 143$ kg mol^{-1}) (number average)

5. $0 \cdot 2$

6. $M_r = 850\,000$ (i.e. $M = 850$ kg mol^{-1}) (mass average)

7. $0 \cdot 8$K of superheating

8.

$c/10^{-3}$ mol dm^{-3}	π/mN m^{-1}	$\Gamma/10^{-6}$ mol m^{-2}	$A/10^{-20}$ m^2
$2 \cdot 0$	$10 \cdot 4$	$2 \cdot 37$	$70 \cdot 2$
$4 \cdot 0$	$20 \cdot 2$	$3 \cdot 24$	$51 \cdot 4$
$6 \cdot 0$	$27 \cdot 1$	$3 \cdot 68$	$45 \cdot 2$

9. (a) 183 mJ m^{-2}
    (b) (i) 970 mJ m^{-2} (ii) 146 mJ m^{-2}
    (c) $+37$ mJ m^{-2}

10. $W_a = 54 \cdot 0$ mJ m^{-2}
    $S \;\; = -91 \cdot 5$ mJ m^{-2}

11. $69°$

12. $M_r = 13\,000$ (suggesting surface dissociation)

13. $V = \dfrac{125\ p/10^5\ \text{N m}^{-2}}{1 + 83\ p/10^5\ \text{N m}^{-2}}\ \text{c}$

14. $15 \cdot 7\ \text{m}^2\ \text{g}^{-1}$

15. $47 \times 10^{-20}\ \text{m}^2$

16. Area $= 64\ \text{m}^2\ \text{g}^{-1}$. Comparison with $42\ \text{m}^2\ \text{g}^{-1}$, as deter-mined by means of an electron microscope, suggests that the carbon black sample is porous

17.
V/V_m	0·4	0·8	1·2
ΔH_{ads}/kJ/mol^{-1}	$-11 \cdot 8$	$-12 \cdot 7$	$-9 \cdot 1$

Values reflect multilayer physical adsorption on a fairly uni-form solid surface

18. $r = 4 \cdot 24\ \text{nm}$ (assuming zero contact angle, cylindrical pore shape, constancy of γ with r, equivalence of multilayer adsorp-tion at flat and curved surfaces)

19. $145 \times 10^{-20}\ \text{m}^2$

20. $\zeta = 32\ \text{mV}$ (using Smoluchowski equation; $\kappa a \approx 130$, there-fore, calculated ζ is probably an underestimate)

21. (a) $u_E = 1 \cdot 5 \times 10^{-8}\ \text{m}^2\ \text{s}^{-1}\ \text{V}^{-1}$
     (b) $3 \cdot 3$ per cent
     (c) $19 \cdot 2\ \text{mV}$ (using Smoluchowski equation; $\kappa a \approx 260$)
     (d) $0 \cdot 014\ \text{C m}^{-2}$

22. $5 \times 10^{-5}\ \text{cm}^3\ \text{s}^{-1}$ towards $-ve$ electrode

23. C (inner part of double layer) $= 0 \cdot 106\ \text{F m}^{-2}$
     $\varepsilon/\varepsilon_0$ (inner part of double layer) $= 4 \cdot 8$ (suggesting orientation of water molecules close to the surface)

24. $k_2 = 7 \cdot 3 \times 10^{-12}$ (particles cm^{-3})$^{-1}$ s^{-1}
     $k_2^0 = \dfrac{4kT}{3\eta} = 6 \cdot 2 \times 10^{-12}$ (particles cm^{-3})$^{-1}$ s^{-1}

25. $M_r = 118\,000$

     M_r (light scattering, mass average) $> M_r$ (viscosity) $> M_r$ (osmotic pressure, number average)

Bibliography

General

[1] McGLASHAN, M. L., *Physico-Chemical Quantities and Units*, R.I.C. Monograph for Teachers, No. 15 (1968).
[2] ADAM, N. K., *The Physics and Chemistry of Surfaces* (3rd ed.), Oxford University Press (1941).
[3] ADAMSON, A. W., *Physical Chemistry of Surfaces* (2nd ed.), Interscience (1967).
[4] ALEXANDER, A. E. and JOHNSON, P., *Colloid Science*, Oxford University Press (1949).
[5] DAVIES, J. T. and RIDEAL, E. K., *Interfacial Phenomena* (2nd ed.), Academic Press (1963).
[6] FLORY, P. J., *Principles of Polymer Chemistry*, Cornell (1953).
[7] GORDON, M., *High Polymers–Structure and Physical Properties* (2nd ed.), Iliffe (1963).
[8] GREGG, S. J., *The Surface Chemistry of Solids* (2nd ed.), Chapman and Hall (1961).
[9] GREGG, S. J. and SING, K. S. W., *Adsorption, Surface Area and Porosity*, Academic Press (1967).
[10] HARKINS, W. D., *The Physical Chemistry of Surface Films*, Reinhold (1952).
[11] KRUYT, H. R. (editor), *Colloid Science*, Elsevier: Volume 1, 'Irreversible Systems' (1952); Volume 2, 'Reversible Systems' (1949).
[12] McBAIN, J. W., *Colloid Science*, Heath (1950).
[13] MOILLIET, J. L., COLLIE, B. and BLACK, W., *Surface Activity (The Physical Chemistry, Technical Applications and Chemical Constitution of Surface Active Agents)* (2nd ed.), Spon (1961).
[14] MYSELS, K. J., *Introduction to Colloid Chemistry*, Interscience (1959).
[15] SHELUDKO, A., *Colloid Chemistry*, Elsevier (1966).

Text

[16] OTTEWILL, R. H. and WOODBRIDGE, R. F., *J. Colloid Sci.*, **16**, 581 (1961).
[17] ZAISER, E. M. and LA MER, V. K., *J. Colloid Sci.*, **3**, 571 (1948).

[18] LAITINEN, H. A., *Chemical Analysis*, McGraw-Hill (1960), Chapters 7–9.
[19] STAUFFER, R. E., in WEISSBERGER, A. (editor), *Technique of Organic Chemistry*, Interscience, 3(1), 65 (1956).
[20] ONCLEY, J. L., in COHN, E. J. and EDSALL, J. T., *Proteins, Amino Acids and Peptides*, A.C.S. Monograph 90, Reinhold (1943).
[21] EINSTEIN, A., *Investigations on the Theory of Brownian Movement*, Methuen (1926); Dover (1956).
[22] GEDDES, A. L., in WEISSBERGER, A. (editor), *Technique of Organic Chemistry*, Interscience, 1(1), 551 (1949).
[23] CADLE, R. D., *Particle Size Determination*, Interscience (1955); *Particle Size—Theory and Industrial Applications*, Reinhold (1965).
[24] ORR, C. and DALLAVALLE, J. M., *Fine Particle Measurement*, Macmillan (1959); ALLEN, T., *Particle Size Measurement*, Chapman and Hall (1968).
[25] SVEDBERG, T. and PEDERSEN, K. O., *The Ultracentrifuge*, Oxford University Press (1940).
[26] NICHOLS, J. B. and BAILEY, E. D., in WEISSBERGER, A. (editor), *Technique of Organic Chemistry*, Interscience, 1(1), 621 (1949).
[27] SCHACHMANN, H. K., *Ultracentrifugation in Biochemistry*, Academic Press (1959).
[28] PICKELS, E. G., *Chem. Rev.*, 30, 351 (1942).
[29] ARCHIBALD, W. J., *J. phys. Chem.*, 51, 1204 (1947).
[30] HUGGINS, M. L., *J. chem. Phys.*, 9, 440 (1941); *J. phys. Chem.*, 46, 151 (1942).
[31] FLORY, P. J., *J. chem. Phys.*, 9, 660 (1941); 10, 51 (1942); 13, 453 (1945).
[32] WAGNER, R. H., in WEISSBERGER, A. (editor), *Technique of Organic Chemistry*, Interscience, 1(1), 487 (1949).
[33] KUPKE, D. W., *Advanc. Protein Chem.*, 15, 57 (1960).
[34] THAIN, J. F., *Principles of Osmotic Phenomena*, R.I.C. Monograph for Teachers, No. 13 (1967).
[35] FUOSS, R. M. and MEAD, D. J., *J. phys. Chem.*, 47, 59 (1943).
[36] STACEY, K. A., *Light-Scattering in Physical Chemistry*, Butterworths (1956).
[37] DEBYE, P., *J. phys, Chem.*, 51, 18 (1947).
[38] VAN DER HULST, H. C., *Light Scattering by Small Particles*, Wiley (1957).
[39] LA MER, V. K. and BARNES, M. D., *J. Colloid Sci.*, 1, 71, 79 (1946).
[40] ZIMM, B. H., *J. chem. Phys.*, 16, 1093, 1099 (1948).
[41] COSSLETT, V. E., *Practical Electron Microscopy*, Butterworths (1951).
[42] WYCKOFF, R. W. G., *The World of the Electron Microscope*, Yale (1958).
[43] KAY, D. (editor), *Techniques for Electron Microscopy*, Blackwell (1961).
[44] OTTEWILL, R. H., 'Electron microscopy of lyophobic colloids', in *The Encyclopedia of Microscopy*, Reinhold (1961).
[45] DE BOER, J. H., *The Dynamical Character of Adsorption* (2nd ed.), Oxford University Press (1968).
[46] HARKINS, W. D. and JORDAN, H. F., *J. Amer. chem. Soc.*, 52, 1751 (1930).
[47] ZUIDEMA, H. H. and WATERS, G. W., *Industr. Engng. Chem. (Anal.)*, 13, 312 (1941).
[48] HARKINS, W. D. and BROWN, F. E., *J. Amer. chem. Soc.*, 41, 499 (1919).
[49] LANDO, J. L. and OAKLEY, H. T., *J. Colloid Interface Sci.*, 25, 526 (1967).
[50] McBAIN, J. W. and SWAIN, R. C., *Proc. roy. Soc.*, A154, 608 (1936).
[51] DIXON, J. K., JUDSON, C. M. and SALLEY, D. J., in SOBOTKA, H. (editor), *Monomolecular Layers*, A.A.A.S. (1954), p. 63.
[52] SCHICK, M. J. (editor), *Non-Ionic Surfactants*, Arnold (1967).
[53] HARTLEY, G. S., *Aqueous Solutions of Paraffin Chain Salts*, Hermann et Cie. (1936).
[54] GARRETT, H. E., Reference 82 (Chapter 2).
[55] ELWORTHY, P. H., FLORENCE, A. T. and MACFARLANE, C. B., *Solubilization by Surface Active Agents*, Chapman and Hall (1968).

[56] PETHICA, B. A., 'Micelle formation', in *Proc. Third int. Congr. Surface Activity, Cologne*, **1**, 212 (1960); HALL, D. G. and PETHICA, B. A., Reference 52 (Chapter 16).

[57] ELWORTHY, P. H. and MYSELS, K. J., *J. Colloid Interface Sci.*, **21**, 331 (1966).

[58] GAINES, G. L., *Insoluble Monolayers at Liquid–Gas Interfaces*, Interscience (1966).

[59] INOKUCHI, K., *Bull. chem. Soc. Japan*, **27**, 203 (1954).

[60] RIES, H. E., 'Monomolecular films', *Sci. Amer.*, March (1961), p. 152.

[61] MARSDEN, J. and RIDEAL, E. K., *J. chem. Soc.*, 1163 (1938).

[62] BULL, H. B., 'Spread monolayers of proteins', *Advanc. Protein Chem.*, **3**, 95 (1947).

[63] CUMPER, C. W. N. and ALEXANDER, A. E., *Trans. Faraday Soc.*, **46**, 235 (1950).

[64] CHEESEMAN, D. F. and DAVIES, J. T., *Advanc. Protein Chem.*, **9**, 439 (1954).

[65] SCHULMAN, J. H. and RIDEAL, E. K., *Proc. roy. Soc.*, **B122**, 46 (1937).

[66] YOUNG, D. M. and CROWELL, A. D., *Physical Adsorption of Gases*, Butterworths (1962).

[67] ROSS, S. and OLIVIER, J. P., *On Physical Adsorption*, Interscience (1964).

[68] HAYWARD, D. O. and TRAPNELL, B. M. W., *Chemisorption* (2nd ed.), Butterworths (1964).

[69] BOND, G. C., *Principles of Catalysis*, R.I.C. Monograph for Teachers, No. 7 (2nd ed.), (1968).

[70] TITOFF, A., *Z. phys. Chem.*, **74**, 641 (1910).

[71] BRUNAUER, S., *Physical Adsorption of Gases and Vapours*, Oxford University Press (1944).

[72] AMBERG, C. H., SPENCER, W. B. and BEEBE, R. A., *Can. J. Chem.*, **33**, 305 (1955).

[73] BRUNAUER, S., EMMETT, P. H. and TELLER, E. J., *J. Amer. chem. Soc.*, **60**, 309 (1938).

[74] DRAIN, L. E., *Sci. Progr.*, **42**, 608 (1954).

[75] BEECK, O., *Discuss. Faraday Soc.*, **7**, 118 (1950).

[76] MCCLELLAN, A. L. and HARNSBERGER, H. F., *J. Colloid Interface Sci.*, **23**, 577 (1967).

[77] LANGMUIR, I. and SCHAEFFER, V., *J. Amer. chem. Soc.*, **59**, 2405 (1937); FORT, T. and PATTERSON, H. T., *J. Colloid Sci.*, **18**, 217 (1963).

[78] JONES, W. C. and PORTER, M. C., *J. Colloid Interface Sci.*, **24**, 1 (1967).

[79] BARTELL, F. E., et al., *J. phys. Chem.*, **36**, 3115 (1932); **38**, 503 (1934).

[80] HARTLEY, G. S., *Chem. & Ind. (Rev.)*, 448 (1960).

[81] GAUDIN, A. M., *Flotation* (2nd ed.), McGraw-Hill (1957).

[82] DURHAM, K. (editor), *Surface Activity and Detergency*, Macmillan (1961).

[83] ADAM, N. K. and STEVENSON, D. G., 'Detergent Action', *Endeavour*, **12**, 25 (1953).

[84] KUSHNER, L. M. and HOFFMAN, J. I., 'Synthetic Detergents', *Sci. Amer.*, October (1951), p. 26.

[85] TACHIBANA, T., YABE, A. and TSUBOMURA, M., *J. Colloid Sci.*, **15**, 278 (1960).

[86] KIPLING, J. J., *Adsorption from Solutions of Non-Electrolytes*, Academic Press (1965).

[87] STOCK, R. and RICE, C. B. F., *Chromatographic Methods* (2nd ed), Chapman and Hall (1967).

[88] BOBBITT, J. M., SCHWARTING, A. E. and GRITTER, R. J., *Introduction to Chromatography*, Reinhold (1968).

[89] PURNELL, H., *Gas Chromatography*, Wiley (1962).

[90] RANDERATH, K., *Thin Layer Chromatography* (2nd ed), Academic Press (1966).

[91] DETERMANN, H., *Gel Chromatography*, Springer-Verlag (1968).

[92] INNES, W. B. and ROWLEY, H. H., *J. phys. Chem.*, **51**, 1176 (1951).

[93] BLACKBURN, A. KIPLING, J. J. and TESTER, D. A., *J. chem. Soc.*, 2373 (1957).

[94] OVERBEEK, J. TH. G., Reference 11 (Volume 1, Chapter 4).

[95] GRAHAME, D. C., *Chem. Rev.*, **41**, 441 (1947).

[96] LYKLEMA, J., *Kolloid Zh.*, **175**, 129 (1961); *Discuss. Faraday Soc.*, **42**, 81 (1966).

[97] OVERBEEK, J. TH. G., Reference 11 (Volume 1, Chapter 5).

[98] OTTEWILL, R. H. and WOODBRIDGE, R. F., *J. Colloid Sci.*, **19**, 606 (1964).

[99] SMITH, A. L., in PARFITT, G. D. (editor), *Dispersions of Powders in Liquids*, Chapter 2, Elsevier (1969).

[100] LEVINE, S. and BELL, G. M., *J. Colloid Sci.*, **17**, 838 (1962); LEVINE, S., MINGINS, J. and BELL, G. M., *J. electroanal. Chem.*, **13**, 280 (1967).

[101] SHAW, D. J., *Electrophoresis*, Academic Press (1969).

[102] TISELIUS, A., *Trans. Faraday Soc.*, **33**, 524 (1937).

[103] HENRY, D. C., *Proc. roy. Soc.*, **A133**, 106 (1931).

[104] BOOTH, F., *Trans. Faraday Soc.*, **44**, 955 (1948).

[105] HENRY, D. C., *Trans. Faraday Soc.*, **44**, 1021 (1948).

[106] GHOSH, B. N., *et al.*, *J. Indian chem. Soc.*, **32**, 31 (1955); **40**, 425 (1963).

[107] OVERBEEK, J. TH. G., *Kolloid chem. Beih.*, **54**, 287 (1943); *Advanc. Colloid Sci.*, **3**, 97 (1950).

[108] BOOTH, F., *Nature (Lond.)*, **161**, 83 (1948); *Proc. roy. Soc.*, **A203**, 514 (1950).

[109] WIERSEMA, P. H., LOEB, A. L. and OVERBEEK, J. TH. G., *J. Colloid Interface Sci.* **22**, 78 (1966).

[110] SHAW, J. N. and OTTEWILL, R. H., *Nature (Lond.)*, **208**, 681 (1965).

[111] LYKLEMA, J. and OVERBEEK, J. TH. G., *J. Colloid Sci.*, **16**, 501 (1961).

[112] STIGTER, D., *J. phys. Chem.*, **68**, 3600 (1964).

[113] HUNTER, R. J., *J. Colloid Interface Sci.*, **22**, 231 (1966).

[114] BIEFER, G. J. and MASON, S. G., *Trans. Faraday Soc.*, **55**, 1239 (1959).

[115] MAZUR, P. and OVERBEEK, J. TH. G., *Recl. trav. chim. Pays-Bas*, **70**, 83 (1951).

[116] OVERBEEK, J. TH. G., Reference 11 (Volume 1, Chapter 8).

[117] DERJAGUIN, B. V. and LANDAU, L., *Acta phys.-chim. URSS*, **14**, 633 (1941).

[118] VERWEY, E. J. W. and OVERBEEK, J. TH. G., *Theory of the Stability of Lyophobic Colloids*, Elsevier (1948).

[119] SCHELUDKO, A., *Proc. Acad. Sci. Amsterdam*, **B65**, 76 (1962).

[120] KITCHENER, J. A., *Recent Progr. Surface Sci.*, **1**, 51 (1964).

[121] OVERBEEK, J. TH. G., Reference 11 (Volume 1, Chapter 6).

[122] REERINK, H. and OVERBEEK, J. TH. G., *Discuss. Faraday Soc.*, **18**, 74 (1954).

[123] HAMAKER, H. C., *Physica*, **4**, 1058 (1937).

[124] DERJAGUIN, B. V., ABRICOSSOVA, I. I. and LIFSHITZ, E. M., *Q. Rev. chem. Soc.*, **10**, 295 (1956); DERJAGUIN, B. V., 'The Force Between Molecules', *Sci. Amer.*, July (1960), page 3.

[125] OVERBEEK, J. TH. G., Reference 11 (Volume 1, Chapter 7).

[126] OTTEWILL, R. H. and SHAW, J. N., *Discuss. Faraday Soc.*, **42**, 154 (1966).

[127] FAIRHURST, D., Thesis, Liverpool Polytechnic (1969).

[128] VAN OLPHEN, H., *An Introduction to Clay Colloid Chemistry*, Interscience (1963).

[129] MATIJEVIČ, E. and OTTEWILL, R. H., *J. Colloid Sci.*, **13**, 242 (1958).

[130] LA MER, V. K. and SMELLIE, R., *J. Colloid Sci.*, **11**, 704 (1956); **13**, 589 (1958).

[131] KRAGH, A. M. and LANGSTON, W. B., *J. Colloid Sci.*, **17**, 101 (1962).

[132] EIRICH, F. R. (editor), *Rheology—Theory and Applications*, Academic Press, four volumes (1956–67).

[133] SHERMAN, P., *J. Pharm., Lond.*, **16**, 1 (1964).

[134] RUTGERS, R., *Rheol. Acta*, **2**, 305 (1962).

[135] GUTH, E. and SIMHA, R., *Kolloid Zh.*, **74**, 266 (1936).

[136] FRISCH, H. L. and SIMHA, R., Reference 130 (Volume 1, Chapter 14) (1956).

[137] CONWAY, B. E. and DOBRY-DUCLAUX, A., Reference 130 (Volume 3, Chapter 3) (1960).

[138] MEYERHOFF, G., *Fortschr. Hochpolymeren Forsch.*, **3**, 59 (1961).

[139] VOET, A., *J. phys. Chem.*, **61**, 301 (1957).

[140] FERRY, J. D., *Viscoelastic Properties of Polymers*, Wiley (1961).

[141] TRELOAR, L., *The Physics of Rubber Elasticity* (2nd ed.), Oxford University Press (1958).

[142] BECHER, P., *Emulsions—Theory and Practice* (2nd ed.), A.C.S. Monograph 162, Reinhold (1965).

[143] SHERMAN, P. (editor), *Emulsion Science*, Academic Press (1968).

[144] VAN DEN TEMPEL, M., *Rec. Trav. chim. Pays-Bas*, **72**, 433 (1953).

[145] SCHULMAN, J. H. and COCKBAIN, E. G., *Trans. Faraday Soc.*, **36**, 651 (1940).

[146] GRIFFIN, W. C., *J. Soc. Cosmetic Chemists*, **1**, 311 (1949).

[147] OTTEWILL, R. H. and SHAW, J. N., *Kolloid Zh.*, **215**, 161 (1967).

[148] KITCHENER, J. A. and COOPER, C. F., *Q. Rev. chem. Soc.*, **13**, 71 (1959).

[149] MYSELS, K. J., SHINODA, K. and FRANKEL, S., *Soap Films—Studies of their Thinning*, Pergamon (1959).

[150] DERJAGUIN, B. V. and TITIJEVSKAYA, A. S., *Proc. Second int. Congr. Surface Activity*, Butterworths, London, **1**, 210 (1957).

[151] OVERBEEK, J. TH. G., *J. phys. Chem.*, **64**, 1178 (1960).

[152] MYSELS, K. J., *J. phys. Chem.*, **68**, 3441 (1964).

[153] VAN DEN TEMPEL, M., *J. Colloid Sci.*, **13**, 125 (1958).

[154] EWERS, W. E. and SUTHERLAND, K. L., *Australian J. Sci. Res.*, **A5**, 697 (1952); SHEARER, L. T. and AKERS, W. W., *J. phys. Chem.*, **62**, 1264 (1958).

[155] ROSS, S. and HAAK, R. M., *J. phys. Chem.*, **62**, 1260 (1958).

Index

Adhesion, 77, 117–118, 125–127, 167–186
Adsorption
  activation energy of, 100–102
  at liquid surfaces, 64–70, 79, 80–97
  energies, 99–101, 109, 111, 112–114
  from solution on to solids, 120–132, 134–135, 176, 184–186
  Gibbs equation, 68–71
  hysteresis, 107–108, 115
  isotherms, 104–106, 108–116, 130–132
  of gases and vapours on solids, 98–116
  of ions, 134–135, 139–142, 146–147, 168, 176
  rate of, 67, 101–102
Amphiphilic, 64–65
Antifoaming agents, 218
Archibald technique, 30
Association colloids, 71–77
Asymmetry, 5–6, 18–19, 39–40, 48, 54, 194–195, 197
Averages, 7–8
Avogadro's constant, 21

Barium sulphate sol, 10
B.E.T. adsorption isotherm equation, 111–112, 114–116
Boltzmann-Poisson distribution, 136–138

Born repulsion, 100, 173
Breaking of emulsions, 211
Brownian motion, 18–22, 44, 54, 150, 167, 177–178

Capillary
  condensation, 57, 105–106, 107–108
  rise, 58–60, 107, 119
  viscometer, 189–190
Cardioid condenser, 53–54
Charge effects in diffusion and sedimentation, 30–31
Chemisorption, 99–102, 105, 109, 113
Chi potential, 143–144
Classification of colloidal systems, 2
Clausius-Clapeyron equation, 100
Coagulation, 174 (see Flocculation)
Coalescence of emulsion droplets, 207–208
Cohesion, 77, 118
Co-ions, 133
Colloidal dispersions
  classification of, 3
  preparation of, 8–12, 14–15, 211–213
Concentric cylinder viscometer, 190–192
Condensed monolayers, 86, 88–89, 92, 208, 210
Conductance, 71, 75
  at surfaces, 160–161, 165, 166

233

Cone and plate viscometer, 192
Contact angles, 59, 60, 61, 107–108, 117–126, 208
Counter-ions, 75, 133, 139, 168
Creep, 200–202
Critical micelle concentration, 72, 76, 77
Crystal growth, 9–12

Dark field microscopy, 53–54
Debye equation, 45–47
Debye-Hückel approximation, 138, 159
Denaturation of proteins, 94–95
Derjaguin-Landau and Verwey-Overbeek theory, 168–177
Detergency, 124–128
Dialysis, 13–15
Dielectric dispersion, 40
Diffuse double layer, 135–139
Diffusion, 21–26, 29–32, 177–178
Discreteness of charge effect, 145–146, 180–181
Disjoining pressure, 215–216
Dispersed phase, 3
Dispersion medium, 3
Dispersions
  classification of, 3
  preparation of, 8–12, 14–15, 211–213
Dissymmetry of scattering, 46–48
Donnan membrane equilibrium, 36–39
Drop volume and drop weight, 62–63
Duplex films, 78, 91, 209
Dupré equation, 77, 117

Einstein
  Brownian displacement equation, 20, 21–22
  diffusion equation, 20, 22–23
  viscosity equation, 193
Electric double layer, 133–146, 156–166, 169, 215–216
Electrodecantation, 15
Electrodialysis, 15
Electrokinetic phenomena, 147–165
Electrokinetic (zeta) potential, 143, 144–145, 156–166
Electron microscopy, 48–52, 85–86
Electro-osmosis, 147, 149, 152, 155–156, 159, 166
Electrophoresis, 29, 54, 134, 147, 148–154, 156–164
Electrophoretic retardation, 161
Electroviscous effects, 195
Emulsifying agents, 207–211

Emulsion polymerisation, 211–213
Emulsions, 3, 206–211
Emulsion type, 208–211
Entropic stabilisation, 185
Evaporation through monolayers, 92–93
Expanded monolayers, 86, 90–91, 92, 208

Fick's laws of diffusion, 21, 25, 29
Flexibility of polymer molecules, 6–7, 20, 195–196, 203–204
Flocculation, 29, 54, 167–186, 196, 198–199, 207
  concentrations, 168, 174–177, 181
  kinetics, 177–181
Flotation, 122–124
Fluctuation theory of light scattering, 45
Foaming agents, 123–124, 126, 207, 213–218
Foams, 3, 4, 213–218
Freundlich adsorption isotherm equation, 111, 130–132
Frictional coefficient, 17–18, 20–23
Frictional ratio, 18–19, 23, 29, 32

Gas adsorption, 98–116
  areas of solids from, 114–116
  gravimetric measurement of, 103–104
  volumetric measurement of, 103
Gaseous monolayers, 87–88, 95–96
Gel permeation chromatography, 154
Gels, 7, 182, 199
Gibbs adsorption equation, 68–71, 74, 87, 134
Gibbs-Marangoni effect, 217
Gold sols, 9, 11
Gouy-Chapman model of diffuse double layer, 135–139, 169
Grahame model of inner part of double layer, 145

Hamaker constant, 170–171, 176
Heats of adsorption, 99–101, 109, 111, 112–114
Henry equation, 159–161
Higher-order Tyndall spectra, 47
HLB, 210–211
Hookean elasticity, 187, 199, 201
Hückel equation, 157, 159
Hydrophilic, 4
Hydrophobic, 4
Hydrosol, 4

Inner part of electric double layer, 139–146
Insoluble surface films, 80–97
Interface, importance of in colloidal systems, 3–4
Interfacial tension, 56, 207–218
Intrinsic viscosity, 193–196
Ion adsorption, 134–135, 139–142, 146–147, 168, 176
Ion dissolution, 135
Ion exchange, 146–147
Iso-electric point, 32, 134, 144–145, 183, 207–208

Kelvin equation, 57, 107
Kinetics of flocculation, 177–181
Krafft phenomenon, 77

Langmuir-Adam surface balance, 81–83
Langmuir adsorption isotherm equation, 108–110, 130–132, 141
Light scattering, 8, 41–48, 178
Lipophilic, 4
Liquid-expanded monolayers, 86, 91
London dispersion forces, 100, 169–171
Lyophilic, 4–5
Lyophobic, 4–5
Lyotropic series, 184

Macromolecules, 6–8, 12–13, 20, 31–33, 44–48, 184–186, 195–196, 198–199, 202–204, 211–213
Mark-Houwink equation, 196
McBain-Bakr sorption balance, 104
Membranes, 14–15, 33–39
Micelles, 71–77, 127–128, 212
Microelectrophoresis, 54, 134, 148–151
Mie theory, 47
Mixed surface films, 96–97, 208
Monodispersed systems, 7, 11–12, 212–213
Monomolecular layers, 64, 80–97, 100, 104–105, 108–110, 129–132
Moving boundary electrophoresis, 29, 151–153
Multilayer adsorption, 100, 104–106, 111–116

Negative adsorption, 66, 129, 131, 134
Newtonian viscosity, 187, 188, 199, 201
Non-linear viscoelasticity, 204
Non-Newtonian flow, 196–199
Nucleation, 9–12, 57–58

Oil well drilling, 182–183
Ore flotation, 122–124
Oriented wedge theory, 209
Origin of surface charge, 133–135
Oscillating jet, 63, 67
Osmometers, 34–37
Osmotic pressure, 8, 31–39, 71

Paint, 183, 199
Particle
    aggregation, 2, 29, 54, 167–186, 196, 198–199, 207
    electrophoresis, 54, 134, 148–151
    shape, 2, 5–6, 18–19, 39–40, 48, 50, 54, 194–195, 197
    size, 1–2, 7–8, 9–12, 13–15, 18, 20, 26, 44–48, 50–52, 54, 73, 177, 181–183, 193, 195–196, 208
Pendant drop, 63
Permittivity in the electric double layer, 137, 142, 163, 177
Physical gas adsorption, 98–116
Plastic flow, 197
Poisson-Boltzmann distribution, 136–138
Polydispersity, 7–8, 11, 25, 29, 30, 34, 46, 54, 196, 209
Polymerisation, 12–13, 211–213
Polymers, 6–8, 12–13, 20, 31–33, 44–48, 184–186, 195–196, 198–199, 202–204, 211–213
Polystyrene latex dispersions, 51, 151, 163, 181, 212–213
Potential-determining ions, 135
Potential energy curves, 101, 171–174
Protective colloids, 184–186
Proteins, 5, 6, 19, 27, 29, 32, 38–39, 93–97, 133–134, 151–154, 183–184, 194, 208

Ramsay-Shields equation, 58
Random coil, 6–7, 20, 195
Rate of adsorption, 67, 217
Rayleigh equation, 43
Relative molecular mass, 7–8, 28–30, 31–34, 44–48, 95–96, 195–196
Relaxation effect in electrophoresis, 161–163
Resolving power, 48
Retarded elasticity, 200
Reversal of charge, 135, 141, 176
Rheology, 187–205
Rheopexy, 199

Ring (du Noüy) tensiometer, 61–62
Rotary Brownian motion, 39–40
Rubber elasticity, 203–204

Salting-out, 184
Schlieren optical method, 24, 26, 152
Schulze-Hardy rule, 168
Secondary minima, 174
Sedimentation
    coefficient, 28
    equilibrium, 20–21, 29–30, 54
    potential, 147
    velocity, 16–17, 26–29
    volume, 181–183
Sensitisation, 185–186
Shear compliance, 201–202
Shear thickening, 198
Shear thinning, 196–199
Silver halide sols, 9, 11–12, 51, 135, 144–145, 168, 180–181
Smoluchowski equation, 157–159
Soaps, 124
Soil, 182
Sols, 3, 8–12, 167–181
Solubilisation, 73–74, 126, 212–213
Solvation, 2, 7, 18–19, 53, 126, 134–135, 183–184, 194–196, 216
Spreading
    coefficient, 78–79
    of liquids on liquids, 78–79
    of liquids on solids, 117–122
Stability
    of emulsions, 207–208
    of foams, 213–218
    of lyophobic sols, 167–183
    of systems containing lyophilic material, 183–186
    ratio, 178–181
Steric stabilisation, 185
Stern layer, 139–146
Stern potential, 139–146, 156, 169, 177
Stokes' law, 17, 157, 193
Streaming
    birefringence, 40, 74
    current, 154–155, 164–165
    potential, 147, 154–155, 164–165
Stress relaxation, 201
Sulphur sols, 9, 11–12, 47
Supersaturation, 9–12, 57–58
Surface active agents (surfactants), 66–67, 121, 124–126, 207–208, 217
Surface
    activity, 64–67

areas of solids, 114–116, 130–132
balance, 81–83
charge density, 138, 140–142, 145
charge, origin of, 133–135
conductance, 160–161, 165, 166
excess concentration, 68–69
film potential, 83–84, 91, 96–97
films of proteins, 93–97, 208
free energy, 55–56, 57
of shear, 143, 147, 163–164
potential, 136, 138–139, 143–144, 156
pressure, 65, 80–83, 87–97
rheology, 85–86, 208, 217–218
tension, 55–67, 69–71, 74, 77–79, 80, 82, 107, 117–121, 125–127, 217
viscosity, 85, 217–218

Thin liquid films, 169, 214–218
Thixotropy, 182–183, 198–199
Translational diffusion, 21–26, 29–32, 178
Traube's rule, 66
Turbidity, 41, 45, 71
Tyndall effect, 41

Ultracentrifuge, 26–30
Ultrafiltration, 15
Ultramicroscope, 53–54

van der Waals forces, 55, 99–100, 118, 141, 169–171, 178
Verwey-Overbeek and Derjaguin-Landau theory, 168–177
Viscoelasticity, 187, 199–205
Viscometers, 189–192
Viscosity, 17, 163–164, 188–199, 208, 217
    coefficient of, 188
    functions of, 192–193
    surface, 85, 217–218

Water repellency, 121–122
Weissenberg effect, 204–205
Wetting, 117–118, 121–122, 124–127, 210–211
Wilhelmy plate, 60–61, 82–83
Work softening, 204

Yield value, 197
Young equation, 117–118, 120
Young-Laplace equation, 57

Zeta potential, 143, 144–145, 156–166, 177
Zimm plot, 47–48
Zone electrophoresis, 153–154